Philosophy of Science: A Very Short Introduction

VERY SHORT INTRODUCTIONS are for anyone wanting a stimulating and accessible way into a new subject. They are written by experts, and have been translated into more than 40 different languages.

The series began in 1995, and now covers a wide variety of topics in every discipline. The VSI library now contains over 450 volumes—a Very Short Introduction to everything from Psychology and Philosophy of Science to American History and Relativity—and continues to grow in every subject area.

Very Short Introductions available now:

ACCOUNTING Christopher Nobes
ADOLESCENCE Peter K. Smith
ADVERTISING Winston Fletcher
AFRICAN AMERICAN RELIGION Eddie S. Glaude Jr
AFRICAN HISTORY John Parker and Richard Rathbone
AFRICAN RELIGIONS Jacob K. Olupona
AGNOSTICISM Robin Le Poidevin
AGRICULTURE Paul Brassley and Richard Soffe
ALEXANDER THE GREAT Hugh Bowden
ALGEBRA Peter M. Higgins
AMERICAN HISTORY Paul S. Boyer
AMERICAN IMMIGRATION David A. Gerber
AMERICAN LEGAL HISTORY G. Edward White
AMERICAN POLITICAL HISTORY Donald Critchlow
AMERICAN POLITICAL PARTIES AND ELECTIONS L. Sandy Maisel
AMERICAN POLITICS Richard M. Valelly
THE AMERICAN PRESIDENCY Charles O. Jones
THE AMERICAN REVOLUTION Robert J. Allison
AMERICAN SLAVERY Heather Andrea Williams
THE AMERICAN WEST Stephen Aron
AMERICAN WOMEN'S HISTORY Susan Ware
ANAESTHESIA Aidan O'Donnell
ANARCHISM Colin Ward
ANCIENT ASSYRIA Karen Radner
ANCIENT EGYPT Ian Shaw
ANCIENT EGYPTIAN ART AND ARCHITECTURE Christina Riggs
ANCIENT GREECE Paul Cartledge
THE ANCIENT NEAR EAST Amanda H. Podany
ANCIENT PHILOSOPHY Julia Annas
ANCIENT WARFARE Harry Sidebottom
ANGELS David Albert Jones
ANGLICANISM Mark Chapman
THE ANGLO-SAXON AGE John Blair
THE ANIMAL KINGDOM Peter Holland
ANIMAL RIGHTS David DeGrazia
THE ANTARCTIC Klaus Dodds
ANTISEMITISM Steven Beller
ANXIETY Daniel Freeman and Jason Freeman
THE APOCRYPHAL GOSPELS Paul Foster
ARCHAEOLOGY Paul Bahn
ARCHITECTURE Andrew Ballantyne
ARISTOCRACY William Doyle
ARISTOTLE Jonathan Barnes
ART HISTORY Dana Arnold
ART THEORY Cynthia Freeland
ASTROBIOLOGY David C. Catling
ASTROPHYSICS James Binney
ATHEISM Julian Baggini
AUGUSTINE Henry Chadwick
AUSTRALIA Kenneth Morgan
AUTISM Uta Frith
THE AVANT GARDE David Cottington
THE AZTECS Davíd Carrasco
BACTERIA Sebastian G. B. Amyes

BARTHES Jonathan Culler
THE BEATS David Sterritt
BEAUTY Roger Scruton
BESTSELLERS John Sutherland
THE BIBLE John Riches
BIBLICAL ARCHAEOLOGY Eric H. Cline
BIOGRAPHY Hermione Lee
BLACK HOLES Katherine Blundell
THE BLUES Elijah Wald
THE BODY Chris Shilling
THE BOOK OF MORMON Terryl Givens
BORDERS Alexander C. Diener and Joshua Hagen
THE BRAIN Michael O'Shea
THE BRICS Andrew F. Cooper
THE BRITISH CONSTITUTION Martin Loughlin
THE BRITISH EMPIRE Ashley Jackson
BRITISH POLITICS Anthony Wright
BUDDHA Michael Carrithers
BUDDHISM Damien Keown
BUDDHIST ETHICS Damien Keown
BYZANTIUM Peter Sarris
CANCER Nicholas James
CAPITALISM James Fulcher
CATHOLICISM Gerald O'Collins
CAUSATION Stephen Mumford and Rani Lill Anjum
THE CELL Terence Allen and Graham Cowling
THE CELTS Barry Cunliffe
CHAOS Leonard Smith
CHEMISTRY Peter Atkins
CHILD PSYCHOLOGY Usha Goswami
CHILDREN'S LITERATURE Kimberley Reynolds
CHINESE LITERATURE Sabina Knight
CHOICE THEORY Michael Allingham
CHRISTIAN ART Beth Williamson
CHRISTIAN ETHICS D. Stephen Long
CHRISTIANITY Linda Woodhead
CITIZENSHIP Richard Bellamy
CIVIL ENGINEERING David Muir Wood
CLASSICAL LITERATURE William Allan
CLASSICAL MYTHOLOGY Helen Morales
CLASSICS Mary Beard and John Henderson
CLAUSEWITZ Michael Howard
CLIMATE Mark Maslin
CLIMATE CHANGE Mark Maslin
THE COLD WAR Robert McMahon
COLONIAL AMERICA Alan Taylor
COLONIAL LATIN AMERICAN LITERATURE Rolena Adorno
COMBINATORICS Robin Wilson
COMEDY Matthew Bevis
COMMUNISM Leslie Holmes
COMPLEXITY John H. Holland
THE COMPUTER Darrel Ince
COMPUTER SCIENCE Subrata Dasgupta
CONFUCIANISM Daniel K. Gardner
THE CONQUISTADORS Matthew Restall and Felipe Fernández-Armesto
CONSCIENCE Paul Strohm
CONSCIOUSNESS Susan Blackmore
CONTEMPORARY ART Julian Stallabrass
CONTEMPORARY FICTION Robert Eaglestone
CONTINENTAL PHILOSOPHY Simon Critchley
CORAL REEFS Charles Sheppard
CORPORATE SOCIAL RESPONSIBILITY Jeremy Moon
CORRUPTION Leslie Holmes
COSMOLOGY Peter Coles
CRIME FICTION Richard Bradford
CRIMINAL JUSTICE Julian V. Roberts
CRITICAL THEORY Stephen Eric Bronner
THE CRUSADES Christopher Tyerman
CRYPTOGRAPHY Fred Piper and Sean Murphy
CRYSTALLOGRAPHY A. M. Glazer
THE CULTURAL REVOLUTION Richard Curt Kraus
DADA AND SURREALISM David Hopkins
DANTE Peter Hainsworth and David Robey
DARWIN Jonathan Howard
THE DEAD SEA SCROLLS Timothy Lim
DECOLONIZATION Dane Kennedy
DEMOCRACY Bernard Crick
DERRIDA Simon Glendinning
DESCARTES Tom Sorell
DESERTS Nick Middleton
DESIGN John Heskett
DEVELOPMENTAL BIOLOGY Lewis Wolpert
THE DEVIL Darren Oldridge
DIASPORA Kevin Kenny
DICTIONARIES Lynda Mugglestone

DINOSAURS David Norman
DIPLOMACY Joseph M. Siracusa
DOCUMENTARY FILM Patricia Aufderheide
DREAMING J. Allan Hobson
DRUGS Les Iversen
DRUIDS Barry Cunliffe
EARLY MUSIC Thomas Forrest Kelly
THE EARTH Martin Redfern
EARTH SYSTEM SCIENCE Tim Lenton
ECONOMICS Partha Dasgupta
EDUCATION Gary Thomas
EGYPTIAN MYTH Geraldine Pinch
EIGHTEENTH-CENTURY BRITAIN Paul Langford
THE ELEMENTS Philip Ball
EMOTION Dylan Evans
EMPIRE Stephen Howe
ENGELS Terrell Carver
ENGINEERING David Blockley
ENGLISH LITERATURE Jonathan Bate
THE ENLIGHTENMENT John Robertson
ENTREPRENEURSHIP Paul Westhead and Mike Wright
ENVIRONMENTAL ECONOMICS Stephen Smith
ENVIRONMENTAL POLITICS Andrew Dobson
EPICUREANISM Catherine Wilson
EPIDEMIOLOGY Rodolfo Saracci
ETHICS Simon Blackburn
ETHNOMUSICOLOGY Timothy Rice
THE ETRUSCANS Christopher Smith
THE EUROPEAN UNION John Pinder and Simon Usherwood
EVOLUTION Brian and Deborah Charlesworth
EXISTENTIALISM Thomas Flynn
EXPLORATION Stewart A. Weaver
THE EYE Michael Land
FAMILY LAW Jonathan Herring
FASCISM Kevin Passmore
FASHION Rebecca Arnold
FEMINISM Margaret Walters
FILM Michael Wood
FILM MUSIC Kathryn Kalinak
THE FIRST WORLD WAR Michael Howard
FOLK MUSIC Mark Slobin
FOOD John Krebs
FORENSIC PSYCHOLOGY David Canter
FORENSIC SCIENCE Jim Fraser
FORESTS Jaboury Ghazoul
FOSSILS Keith Thomson
FOUCAULT Gary Gutting
THE FOUNDING FATHERS R. B. Bernstein
FRACTALS Kenneth Falconer
FREE SPEECH Nigel Warburton
FREE WILL Thomas Pink
FRENCH LITERATURE John D. Lyons
THE FRENCH REVOLUTION William Doyle
FREUD Anthony Storr
FUNDAMENTALISM Malise Ruthven
FUNGI Nicholas P. Money
GALAXIES John Gribbin
GALILEO Stillman Drake
GAME THEORY Ken Binmore
GANDHI Bhikhu Parekh
GENES Jonathan Slack
GENIUS Andrew Robinson
GEOGRAPHY John Matthews and David Herbert
GEOPOLITICS Klaus Dodds
GERMAN LITERATURE Nicholas Boyle
GERMAN PHILOSOPHY Andrew Bowie
GLOBAL CATASTROPHES Bill McGuire
GLOBAL ECONOMIC HISTORY Robert C. Allen
GLOBALIZATION Manfred Steger
GOD John Bowker
GOETHE Ritchie Robertson
THE GOTHIC Nick Groom
GOVERNANCE Mark Bevir
THE GREAT DEPRESSION AND THE NEW DEAL Eric Rauchway
HABERMAS James Gordon Finlayson
HAPPINESS Daniel M. Haybron
THE HARLEM RENAISSANCE Cheryl A. Wall
THE HEBREW BIBLE AS LITERATURE Tod Linafelt
HEGEL Peter Singer
HEIDEGGER Michael Inwood
HERMENEUTICS Jens Zimmermann
HERODOTUS Jennifer T. Roberts
HIEROGLYPHS Penelope Wilson
HINDUISM Kim Knott
HISTORY John H. Arnold

THE HISTORY OF ASTRONOMY Michael Hoskin
THE HISTORY OF CHEMISTRY William H. Brock
THE HISTORY OF LIFE Michael Benton
THE HISTORY OF MATHEMATICS Jacqueline Stedall
THE HISTORY OF MEDICINE William Bynum
THE HISTORY OF TIME Leofranc Holford-Strevens
HIV/AIDS Alan Whiteside
HOBBES Richard Tuck
HOLLYWOOD Peter Decherney
HORMONES Martin Luck
HUMAN ANATOMY Leslie Klenerman
HUMAN EVOLUTION Bernard Wood
HUMAN RIGHTS Andrew Clapham
HUMANISM Stephen Law
HUME A. J. Ayer
HUMOUR Noël Carroll
THE ICE AGE Jamie Woodward
IDEOLOGY Michael Freeden
INDIAN PHILOSOPHY Sue Hamilton
INFECTIOUS DISEASE Marta L. Wayne and Benjamin M. Bolker
INFORMATION Luciano Floridi
INNOVATION Mark Dodgson and David Gann
INTELLIGENCE Ian J. Deary
INTERNATIONAL LAW Vaughan Lowe
INTERNATIONAL MIGRATION Khalid Koser
INTERNATIONAL RELATIONS Paul Wilkinson
INTERNATIONAL SECURITY Christopher S. Browning
IRAN Ali M. Ansari
ISLAM Malise Ruthven
ISLAMIC HISTORY Adam Silverstein
ISOTOPES Rob Ellam
ITALIAN LITERATURE Peter Hainsworth and David Robey
JESUS Richard Bauckham
JOURNALISM Ian Hargreaves
JUDAISM Norman Solomon
JUNG Anthony Stevens
KABBALAH Joseph Dan
KAFKA Ritchie Robertson
KANT Roger Scruton
KEYNES Robert Skidelsky
KIERKEGAARD Patrick Gardiner
KNOWLEDGE Jennifer Nagel
THE KORAN Michael Cook
LANDSCAPE ARCHITECTURE Ian H. Thompson
LANDSCAPES AND GEOMORPHOLOGY Andrew Goudie and Heather Viles
LANGUAGES Stephen R. Anderson
LATE ANTIQUITY Gillian Clark
LAW Raymond Wacks
THE LAWS OF THERMODYNAMICS Peter Atkins
LEADERSHIP Keith Grint
LEARNING Mark Haselgrove
LIBERALISM Michael Freeden
LIGHT Ian Walmsley
LINCOLN Allen C. Guelzo
LINGUISTICS Peter Matthews
LITERARY THEORY Jonathan Culler
LOCKE John Dunn
LOGIC Graham Priest
LOVE Ronald de Sousa
MACHIAVELLI Quentin Skinner
MADNESS Andrew Scull
MAGIC Owen Davies
MAGNA CARTA Nicholas Vincent
MAGNETISM Stephen Blundell
MALTHUS Donald Winch
MANAGEMENT John Hendry
MAO Delia Davin
MARINE BIOLOGY Philip V. Mladenov
THE MARQUIS DE SADE John Phillips
MARTIN LUTHER Scott H. Hendrix
MARTYRDOM Jolyon Mitchell
MARX Peter Singer
MATERIALS Christopher Hall
MATHEMATICS Timothy Gowers
THE MEANING OF LIFE Terry Eagleton
MEDICAL ETHICS Tony Hope
MEDICAL LAW Charles Foster
MEDIEVAL BRITAIN John Gillingham and Ralph A. Griffiths
MEDIEVAL LITERATURE Elaine Treharne
MEDIEVAL PHILOSOPHY John Marenbon
MEMORY Jonathan K. Foster

METAPHYSICS Stephen Mumford
THE MEXICAN REVOLUTION
Alan Knight
MICHAEL FARADAY
Frank A. J. L. James
MICROBIOLOGY Nicholas P. Money
MICROECONOMICS Avinash Dixit
MICROSCOPY Terence Allen
THE MIDDLE AGES Miri Rubin
MINERALS David Vaughan
MODERN ART David Cottington
MODERN CHINA Rana Mitter
MODERN DRAMA
Kirsten E. Shepherd-Barr
MODERN FRANCE Vanessa R. Schwartz
MODERN IRELAND Senia Pašeta
MODERN JAPAN
Christopher Goto-Jones
MODERN LATIN AMERICAN
LITERATURE
Roberto González Echevarría
MODERN WAR Richard English
MODERNISM Christopher Butler
MOLECULES Philip Ball
THE MONGOLS Morris Rossabi
MOONS David A. Rothery
MORMONISM Richard Lyman Bushman
MOUNTAINS Martin F. Price
MUHAMMAD Jonathan A. C. Brown
MULTICULTURALISM Ali Rattansi
MUSIC Nicholas Cook
MYTH Robert A. Segal
THE NAPOLEONIC WARS
Mike Rapport
NATIONALISM Steven Grosby
NELSON MANDELA Elleke Boehmer
NEOLIBERALISM Manfred Steger and
Ravi Roy
NETWORKS Guido Caldarelli and
Michele Catanzaro
THE NEW TESTAMENT
Luke Timothy Johnson
THE NEW TESTAMENT AS
LITERATURE Kyle Keefer
NEWTON Robert Iliffe
NIETZSCHE Michael Tanner
NINETEENTH-CENTURY BRITAIN
Christopher Harvie and
H. C. G. Matthew
THE NORMAN CONQUEST
George Garnett
NORTH AMERICAN INDIANS
Theda Perdue and Michael D. Green
NORTHERN IRELAND
Marc Mulholland
NOTHING Frank Close
NUCLEAR PHYSICS Frank Close
NUCLEAR POWER Maxwell Irvine
NUCLEAR WEAPONS
Joseph M. Siracusa
NUMBERS Peter M. Higgins
NUTRITION David A. Bender
OBJECTIVITY Stephen Gaukroger
THE OLD TESTAMENT
Michael D. Coogan
THE ORCHESTRA D. Kern Holoman
ORGANIZATIONS Mary Jo Hatch
PAGANISM Owen Davies
THE PALESTINIAN-ISRAELI CONFLICT
Martin Bunton
PARTICLE PHYSICS Frank Close
PAUL E. P. Sanders
PEACE Oliver P. Richmond
PENTECOSTALISM William K. Kay
THE PERIODIC TABLE Eric R. Scerri
PHILOSOPHY Edward Craig
PHILOSOPHY IN THE ISLAMIC
WORLD Peter Adamson
PHILOSOPHY OF LAW
Raymond Wacks
PHILOSOPHY OF SCIENCE
Samir Okasha
PHOTOGRAPHY Steve Edwards
PHYSICAL CHEMISTRY Peter Atkins
PILGRIMAGE Ian Reader
PLAGUE Paul Slack
PLANETS David A. Rothery
PLANTS Timothy Walker
PLATE TECTONICS Peter Molnar
PLATO Julia Annas
POLITICAL PHILOSOPHY David Miller
POLITICS Kenneth Minogue
POSTCOLONIALISM Robert Young
POSTMODERNISM Christopher Butler
POSTSTRUCTURALISM
Catherine Belsey
PREHISTORY Chris Gosden
PRESOCRATIC PHILOSOPHY
Catherine Osborne

PRIVACY Raymond Wacks
PROBABILITY John Haigh
PROGRESSIVISM Walter Nugent
PROTESTANTISM Mark A. Noll
PSYCHIATRY Tom Burns
PSYCHOANALYSIS Daniel Pick
PSYCHOLOGY Gillian Butler and Freda McManus
PSYCHOTHERAPY Tom Burns and Eva Burns-Lundgren
PURITANISM Francis J. Bremer
THE QUAKERS Pink Dandelion
QUANTUM THEORY John Polkinghorne
RACISM Ali Rattansi
RADIOACTIVITY Claudio Tuniz
RASTAFARI Ennis B. Edmonds
THE REAGAN REVOLUTION Gil Troy
REALITY Jan Westerhoff
THE REFORMATION Peter Marshall
RELATIVITY Russell Stannard
RELIGION IN AMERICA Timothy Beal
THE RENAISSANCE Jerry Brotton
RENAISSANCE ART Geraldine A. Johnson
REVOLUTIONS Jack A. Goldstone
RHETORIC Richard Toye
RISK Baruch Fischhoff and John Kadvany
RITUAL Barry Stephenson
RIVERS Nick Middleton
ROBOTICS Alan Winfield
ROMAN BRITAIN Peter Salway
THE ROMAN EMPIRE Christopher Kelly
THE ROMAN REPUBLIC David M. Gwynn
ROMANTICISM Michael Ferber
ROUSSEAU Robert Wokler
RUSSELL A. C. Grayling
RUSSIAN HISTORY Geoffrey Hosking
RUSSIAN LITERATURE Catriona Kelly
THE RUSSIAN REVOLUTION S. A. Smith
SAVANNA Peter A. Furley
SCHIZOPHRENIA Chris Frith and Eve Johnstone
SCHOPENHAUER Christopher Janaway
SCIENCE AND RELIGION Thomas Dixon
SCIENCE FICTION David Seed
THE SCIENTIFIC REVOLUTION Lawrence M. Principe
SCOTLAND Rab Houston
SEXUALITY Véronique Mottier
SHAKESPEARE'S COMEDIES Bart van Es
SIKHISM Eleanor Nesbitt
THE SILK ROAD James A. Millward
SLANG Jonathon Green
SLEEP Steven W. Lockley and Russell G. Foster
SOCIAL AND CULTURAL ANTHROPOLOGY John Monaghan and Peter Just
SOCIAL PSYCHOLOGY Richard J. Crisp
SOCIAL WORK Sally Holland and Jonathan Scourfield
SOCIALISM Michael Newman
SOCIOLINGUISTICS John Edwards
SOCIOLOGY Steve Bruce
SOCRATES C. C. W. Taylor
SOUND Mike Goldsmith
THE SOVIET UNION Stephen Lovell
THE SPANISH CIVIL WAR Helen Graham
SPANISH LITERATURE Jo Labanyi
SPINOZA Roger Scruton
SPIRITUALITY Philip Sheldrake
SPORT Mike Cronin
STARS Andrew King
STATISTICS David J. Hand
STEM CELLS Jonathan Slack
STRUCTURAL ENGINEERING David Blockley
STUART BRITAIN John Morrill
SUPERCONDUCTIVITY Stephen Blundell
SYMMETRY Ian Stewart
TAXATION Stephen Smith
TEETH Peter S. Ungar
TERRORISM Charles Townshend
THEATRE Marvin Carlson
THEOLOGY David F. Ford
THOMAS AQUINAS Fergus Kerr
THOUGHT Tim Bayne
TIBETAN BUDDHISM Matthew T. Kapstein
TOCQUEVILLE Harvey C. Mansfield
TRAGEDY Adrian Poole
THE TROJAN WAR Eric H. Cline
TRUST Katherine Hawley
THE TUDORS John Guy
TWENTIETH-CENTURY BRITAIN Kenneth O. Morgan

THE UNITED NATIONS Jussi M. Hanhimäki
THE U.S. CONGRESS Donald A. Ritchie
THE U.S. SUPREME COURT Linda Greenhouse
UTOPIANISM Lyman Tower Sargent
THE VIKINGS Julian Richards
VIRUSES Dorothy H. Crawford
WATER John Finney
THE WELFARE STATE David Garland
WILLIAM SHAKESPEARE Stanley Wells
WITCHCRAFT Malcolm Gaskill
WITTGENSTEIN A. C. Grayling
WORK Stephen Fineman
WORLD MUSIC Philip Bohlman
THE WORLD TRADE ORGANIZATION Amrita Narlikar
WORLD WAR II Gerhard L. Weinberg
WRITING AND SCRIPT Andrew Robinson

Available soon:

BABYLONIA Trevor Bryce
BLOOD Chris Cooper
TRANSLATION Matthew Reynolds
PUBLIC HEALTH Virginia Berridge
INDIAN CINEMA Ashish Rajadhyaksha

For more information visit our website

www.oup.com/vsi/

Samir Okasha

PHILOSOPHY OF SCIENCE

A Very Short Introduction

SECOND EDITION

Great Clarendon Street, Oxford, OX2 6DP,
United Kingdom

Oxford University Press is a department of the University of Oxford.
It furthers the University's objective of excellence in research, scholarship,
and education by publishing worldwide. Oxford is a registered trade mark of
Oxford University Press in the UK and in certain other countries

First edition published 2002
Second edition published 2016

Impression: 14

Published in the United States of America by Oxford University Press
198 Madison Avenue, New York, NY 10016, United States of America

British Library Cataloguing in Publication Data
Data available

Library of Congress Control Number: 2016931583

ISBN 978–0–19–874558–7

Printed in Great Britain by
Ashford Colour Press Ltd, Gosport, Hampshire

For Solomon and Joel

Contents

Acknowledgements

I would like to thank Bill Newton-Smith, Philip Kitcher, Elisabeth Okasha, Shelley Cox, and OUP's readers for their comments and suggestions. Thanks also to the numerous readers of the first edition who sent me their thoughts, criticisms, and comments, which I have tried to take into account in preparing the second edition.

List of illustrations

Chapter 1
What is science?

What is science? This question may seem easy to answer: everybody knows that subjects such as physics, chemistry, and biology constitute science, while subjects such as art, music, and theology do not. But when as philosophers we ask what science is, that is not the sort of answer we want. We are not asking for a mere list of the activities that are usually called 'science'. Rather we are asking what common feature all the things on that list share, i.e. what it is that *makes* something a science. Understood this way, our question is not so trivial.

But you may still think the question is relatively straightforward. Surely science is just the attempt to understand, explain, and predict the world we live in? This is certainly a reasonable answer. But is it the whole story? After all, the various religions also attempt to understand and explain the world, but religion is not usually regarded as a branch of science. Similarly, astrology and fortune-telling are attempts to predict the future, but most people would not describe these activities as science. Or consider history. Historians try to understand and explain what happened in the past, but history is usually classified as a humanities subject not a science subject. As with many philosophical questions, the question 'what is science?' is trickier than it looks at first sight.

Many people believe that the distinguishing features of science lie in the particular *methods* scientists use to investigate the world. This suggestion is quite plausible. For many scientific disciplines do employ distinctive methods of enquiry that are not used in non-scientific enterprises. An obvious example is the use of experiments, which historically marks a turning-point in the development of modern science. Not all the sciences are experimental though—astronomers obviously cannot do experiments on the heavens, but have to content themselves with careful observation instead. The same is true of many social sciences. Another important feature of science is the construction of theories. Scientists do not simply record the results of experiment and observation in a log book—they usually want to explain those results in terms of a general theory. This is not always easy to do, but there have been some striking successes. One of the main tasks of philosophy of science is to understand how techniques such as experimentation, observation, and theory construction have enabled scientists to unravel so many of nature's secrets.

The origins of modern science

In today's schools and universities, science is taught in a largely ahistorical way. Textbooks present the key ideas of a scientific discipline in as convenient a form as possible, with little mention of the lengthy and often tortuous historical process which led to their discovery. As a pedagogical strategy, this makes good sense. But some appreciation of the history of scientific ideas is helpful for understanding the issues that interest philosophers of science. Indeed as we shall see in Chapter 5, it has been argued that close attention to the history of science is indispensable for doing good philosophy of science.

The origins of modern science lie in a period of rapid scientific development that occurred in Europe between about 1500 and 1750, which we now refer to as the *scientific revolution*. Of course scientific investigations were pursued in ancient and medieval times too—the scientific revolution did not come from nowhere. In these earlier periods the dominant worldview was *Aristotelianism*, named after the ancient Greek philosopher Aristotle, who put forward detailed theories in physics, biology, astronomy, and cosmology. But Aristotle's ideas would seem very strange to a modern scientist, as would his methods of enquiry. To pick just one example, he believed that all earthly bodies are composed of just four elements: earth, fire, air, and water. This view is obviously at odds with what modern chemistry tells us.

The first crucial step in the development of the modern scientific worldview was the Copernican revolution. In 1542 the Polish astronomer Nicolas Copernicus (1473–1543) published a book attacking the geocentric model of the universe, which placed the stationary earth at the centre of the universe with the planets and the sun in orbit around it. Geocentric astronomy, also known as Ptolemaic astronomy after the ancient Greek astronomer Ptolemy, lay at the heart of the Aristotelian worldview, and had gone largely unchallenged for 1,800 years. But Copernicus suggested an alternative: the *sun* was the fixed centre of the universe, and the planets, including the earth, were in orbit around it (see Figure 1). On this heliocentric model the earth is regarded as just another planet, and so loses the unique status that tradition had accorded it. Copernicus' theory initially met with much resistance, not least from the Catholic Church who regarded it as contravening the Scriptures, and in 1616 banned books advocating the earth's motion. But within 100 years Copernicanism had become established scientific orthodoxy.

Copernicus' innovation did not merely lead to a better astronomy. Indirectly, it led to the development of modern physics, through the work of Johannes Kepler (1571–1630) and Galileo Galilei (1564–1642). Kepler discovered that the planets do not move in circular orbits around the sun, as Copernicus thought, but rather in *ellipses*. This was his 'first law' of planetary motion; his second

1. Copernicus' heliocentric model of the universe, showing the planets, including the earth, orbiting the sun.

and third laws specify the speeds at which the planets orbit the sun. Taken together, Kepler's laws provided a successful planetary theory, solving problems that had confounded astronomers for centuries. Galileo was a lifelong supporter of Copernicanism and one of the early pioneers of the telescope. When he pointed his telescope at the heavens, he made a wealth of amazing discoveries: mountains on the moon, a vast array of stars, sun-spots, Jupiter's moons, and more. All of these conflicted with Aristotelian cosmology, and played a pivotal role in converting the scientific community to Copernicanism.

Galileo's most enduring contribution, however, lay not in astronomy but in mechanics, where he refuted the Aristotelian theory that heavier bodies fall faster than lighter ones. In place of this theory, Galileo made the counter-intuitive suggestion that all freely falling bodies will fall towards the earth at the same rate, irrespective of their weight. (Of course in practice, if you drop a feather and a cannonball from the same height the cannonball will land first, but Galileo argued that this is simply due to air resistance—in a vacuum, they would land together.) Furthermore, he argued that freely falling bodies accelerate uniformly, i.e. gain equal increments of speed in equal times; this is known as Galileo's law of free fall. Galileo provided persuasive though not conclusive evidence for this law, which formed the centrepiece of his mechanics.

Galileo is generally regarded as the first modern physicist. He was the first to show that the language of mathematics could be used to describe the behaviour of material objects, such as falling bodies and projectiles. To us this seems obvious—today's scientific theories are routinely formulated in mathematical language, not only in physics but also in the biological and social sciences. But in Galileo's day it was not obvious: mathematics was widely regarded as dealing with purely abstract entities, hence inapplicable to physical reality. Another innovative aspect was Galileo's emphasis on testing hypotheses experimentally. To the modern scientist this may again seem obvious. But in Galileo's day experimentation was not generally regarded as a reliable means of gaining knowledge. Galileo's emphasis on experiment marks the beginning of an empirical approach to studying nature that continues to this day.

The period following Galileo's death saw the scientific revolution rapidly gain in momentum. The French philosopher-scientist René Descartes (1596–1650) developed a radical new 'mechanical philosophy', according to which the physical world consists of inert particles of matter interacting and colliding with one another. The laws governing the motion of these particles or 'corpuscles' held the key to understanding the structure of the universe, Descartes believed. The mechanical philosophy promised to explain all observable phenomena in terms of the motions of these corpuscles, and quickly became the dominant scientific vision of the late 17th century; to some extent it is still with us today. Versions of the mechanical philosophy were espoused by figures such as Huygens, Gassendi, Hooke, and Boyle; its acceptance marked the final downfall of the Aristotelian worldview.

The scientific revolution culminated in the work of Isaac Newton (1643–1727), whose masterpiece, *Mathematical Principles of Natural Philosophy*, was published in 1687.

Newton agreed with the mechanical philosophers that the universe consists simply of particles in motion, and sought to improve on Descartes's theory. The result was a dynamical and mechanical theory of great power, based around Newton's three laws of motion and his famous principle of *universal gravitation*. According to this principle, every body in the universe exerts a gravitational attraction on every other body; the strength of the attraction between two bodies depends on the product of their masses, and on the distance between them squared. The laws of motion then specify how this gravitational force affects the bodies' motions. Newton elaborated his theory with remarkable precision and rigour, inventing the mathematical techniques we now call 'calculus'. Strikingly, Newton was able to show that Kepler's laws of planetary motion and Galileo's law of free fall (both with certain minor modifications) were logical consequences of his laws of motion and gravitation. So a single set of laws could explain the motions of bodies in both terrestrial and celestial domains, and were formulated by Newton in a precise quantitative form.

Newtonian physics provided the framework for science for the next 200 years, quickly replacing Cartesian physics. Scientific confidence grew rapidly in this period, due largely to the success of Newton's theory, which was widely believed to have revealed the true workings of nature, and to be capable of explaining everything, in principle at least. Detailed attempts were made to extend the Newtonian mode of explanation to more and more phenomena. The 18th and 19th centuries both saw notable scientific advances, particularly in chemistry, optics, thermodynamics, and electromagnetism. But for the most part, these developments were regarded as falling within a broadly Newtonian conception of the universe. Scientists accepted Newton's conception as essentially correct; what remained to be done was to fill in the details.

Confidence in the Newtonian picture was shattered in the early years of the 20th century, thanks to two revolutionary new developments in physics: relativity theory and quantum mechanics. Relativity theory, discovered by Einstein, showed that Newtonian mechanics does not give the right results when applied to very massive objects, or objects moving at very high velocities. Quantum mechanics, conversely, shows that the Newtonian theory does not work when applied on a very small scale, to subatomic particles. Both relativity theory and quantum mechanics, especially the latter, are strange and radical theories, making claims about the nature of reality which conflict with common sense, and which many people find hard to accept or even understand. Their emergence caused considerable conceptual upheaval in physics, which continues to this day.

So far our brief account of the history of science has focused mainly on physics. This is no accident, as physics is both historically important and in a sense the most fundamental scientific discipline. For the objects that other sciences study are themselves made up of physical entities, but not vice versa. Consider botany, for example. Botanists study plants, which are composed of cells, which are themselves composed of bio-molecules, which are ultimately made up of atoms, which are physical particles. So botany deals with entities that are less 'fundamental' than does physics—though that is not to say it is less important. This is a point we shall return to in Chapter 3. But even a brief description of modern science's origins would be incomplete if it omitted all mention of the non-physical sciences.

In biology, the event that stands out is Charles Darwin's discovery of the theory of evolution by natural selection, published in *The Origin of Species* in 1859. Until then it was widely believed that the different species had been separately created by God, as the Book of Genesis teaches. But Darwin argued that contemporary species have actually evolved from ancestral ones, through a process known as *natural selection*. Natural selection occurs when some organisms leave more offspring than others, depending on their physical characteristics; if these characteristics are then inherited by their offspring, over time the population will become better and better adapted to the environment. Simple though this process is, over a large number of generations it can cause one species to evolve into a wholly new one, Darwin argued. So persuasive was the evidence Darwin adduced for his theory that by the start of the 20th century it was accepted as scientific orthodoxy, despite considerable theological opposition. Subsequent work has provided striking confirmation of Darwin's theory, which forms the centrepiece of the modern biological worldview.

The 20th century witnessed another revolution in biology that is not yet complete: the emergence of molecular biology and genetics. In 1953 Watson and Crick discovered the structure of DNA, the hereditary material that makes up the genes in the cells of living creatures (see Figure 2). Watson and Crick's discovery explained how genetic information can be copied from one cell to another, and thus passed down from parent to offspring, thereby explaining why offspring tend to resemble their parents. Their discovery opened up an exciting new area of biological research known as molecular biology, which studies the molecular basis of biological phenomena. In the sixty years since Watson and Crick's work, molecular biology has grown fast, transforming our understanding of heredity, development, and other core biological processes. In 2003, the decade-long attempt to provide a molecular-level description of the complete set of genes in a human being, known as the Human Genome Project, was finally

2. James Watson and Francis Crick with the famous 'double helix'—their molecular model of the structure of DNA, discovered in 1953.

completed; the implications for medicine and biotechnology have only begun to be explored. The 21st century will most likely see further exciting developments in this field.

More resources have been devoted to scientific research in the last sixty years than ever before. One result has been an explosion of new scientific disciplines, such as computer science, artificial intelligence, and neuroscience. The late 20th century witnessed the rise of cognitive science, which studies aspects of human cognition including perception, memory, and reasoning, and has transformed traditional psychology. Much of the impetus for cognitive science comes from the idea that the human mind is in some respects similar to a computer, and that human mental processes can be understood by comparing them to the operations computers carry out. By contrast, the field of neuroscience studies how the brain itself works. Thanks to technological advances in brain scanning, neuroscientists are beginning to understand the underlying neural basis of human (and animal) cognition. This enterprise is of great intrinsic interest and may also lead to improved treatments for mental disorders.

The social sciences, such as economics, anthropology, and sociology, also flourished in the 20th century, though some believe they lag behind the natural sciences in terms of sophistication and predictive power. This raises an interesting methodological question. Should social scientists try to use the same methods as natural scientists, or does their subject matter call for a different approach? We return to this issue in Chapter 7.

What is philosophy of science?

The principal task of philosophy of science is to analyse the methods of enquiry used in the sciences. You may wonder why this task should fall to philosophers, rather than to the scientists themselves. This is a good question. Part of the answer is that philosophical reflection can uncover assumptions that are implicit in scientific enquiry. To illustrate, consider experimental practice. Suppose a scientist does an experiment and gets a particular result. They repeat the experiment a few times and keep getting the same result. After that they will probably stop, confident that were the experiment repeated again under exactly the same conditions, the same result would obtain. This assumption may seem obvious, but as philosophers we want to question it. *Why* assume that future repetitions of the experiment will yield the same result? How do we know this is true? The scientist is unlikely to spend much time puzzling over this: they probably have better things to do. It is a quintessentially philosophical question.

So part of the job of philosophy of science is to question assumptions that scientists take for granted. But it would be wrong to imply that scientists never discuss philosophical issues themselves. Indeed historically, scientists have played a key role in the development of philosophy of science. Descartes, Newton, and Einstein are prominent examples. Each was deeply interested in questions about how science should proceed, what methods of enquiry it should use, and whether there are limits to scientific knowledge. These questions still lie at the heart of contemporary philosophy of science. So the issues that concern philosophers of science have engaged the attention of some of the greatest scientists. That being said, it must be admitted that many scientists today take little interest in philosophy of science, and know little about it. While this is unfortunate, it is not an indication that philosophical issues are no longer relevant. Rather it is a consequence of the increasingly specialized nature of science, and of the polarization between the sciences and the humanities that characterizes much modern education.

You may still be wondering exactly what philosophy of science is all about. For to say that it 'studies the methods of science' is not really to say very much. Rather than try to provide a more informative definition, we will instead examine a classic issue in the philosophy of science.

Science and pseudo-science

Recall the question with which we began: what is science? Karl Popper, an influential 20th-century philosopher of science, thought that the fundamental feature of a scientific theory is that it should be *falsifiable*. To call a theory falsifiable is not to say that it is false. Rather, it means that the theory makes some definite predictions which are capable of being tested against experience. If these predictions turn out to be wrong, then the theory has been falsified, or disproved. So a falsifiable theory is one which we *might* discover to be false—it is not compatible with every possible course of experience. Popper thought that some supposedly scientific theories did not satisfy this condition and thus did not deserve to be called science at all; they were merely pseudo-science.

Freud's psychoanalytic theory was one of Popper's favourite examples of pseudo-science. According to Popper, Freud's theory could be reconciled with any empirical findings whatsoever. Whatever a patient's behaviour, Freudians could find an explanation of it in terms of their theory—they would never admit that their theory was wrong. Popper illustrated his point with the following example. Imagine a man who pushes a child into a river with the intention of murdering him, and another man who sacrifices his life in order to save the child. Freudians can explain both men's behaviour with equal ease: the first was repressed, and the second had achieved sublimation. Popper argued that through the use of such concepts as repression, sublimation, and unconscious desires, Freud's theory could be rendered compatible with any clinical data whatever; it was thus unfalsifiable.

The same was true of Marx's theory of history, Popper maintained. Marx claimed that in industrialized societies around the world, capitalism would give way to socialism and ultimately to communism. But when this didn't happen, instead of admitting that Marx's theory was wrong, Marxists would invent an ad hoc explanation for why what had happened was actually perfectly consistent with their theory. For example, they might say that the inevitable progress to communism had been temporarily slowed by the rise of the welfare state, which 'softened' the proletariat and weakened their revolutionary zeal. In this way, Marx's theory could be made compatible with any possible course of events, just like Freud's. Therefore neither theory qualifies as genuinely scientific, according to Popper's criterion.

Popper contrasted Freud's and Marx's theories with Einstein's theory of gravitation, known as *general relativity*. Unlike Freud's and Marx's theories, Einstein's theory made a very definite prediction: that light rays from distant stars would be deflected by the gravitational field of the sun. Normally this effect would be impossible to observe—except during a solar eclipse. In 1919 the English astrophysicist Sir Arthur Eddington organized two expeditions to observe the solar eclipse of that year, one to Brazil and one to the island of Principe off the Atlantic coast of Africa, with the aim of testing Einstein's prediction. The expeditions found that starlight was indeed deflected by the sun, by almost exactly the amount Einstein had predicted. Popper was very impressed by this. Einstein's theory had made a definite, precise prediction, which was confirmed by observations. Had it turned out that starlight was *not* deflected by the sun, this would have shown that Einstein was wrong. So Einstein's theory satisfies the criterion of falsifiability.

Popper's attempt to demarcate science from pseudo-science is intuitively quite plausible. There is surely something suspicious about a theory that can be made to fit any empirical data whatsoever. But many philosophers regard Popper's criterion as overly simplistic. Popper criticized Freudians and Marxists for explaining away any data which appeared to conflict with their theories, rather than accepting that the theories had been refuted. This certainly looks like a dubious procedure. However there is some evidence that this very procedure is routinely used by 'respectable' scientists—whom Popper would not want to accuse of engaging in pseudo-science—and has led to important scientific discoveries.

Another astronomical example can illustrate this. Newton's gravitational theory, which we encountered earlier, made predictions about the paths the planets should follow as they orbit the sun. For the most part these predictions were borne out by observation. However, the observed orbit of Uranus consistently differed from what Newton's theory predicted. This puzzle was solved in 1846 by two scientists, Adams in England and Leverrier in France, working independently. They suggested that there was another planet, as yet undiscovered, exerting an additional gravitational force on Uranus. Adams and Leverrier were able to calculate the mass and position that this planet would have to have if its gravitational pull was indeed responsible for Uranus' strange behaviour. Shortly afterwards the planet Neptune was discovered, almost exactly where Adams and Leverrier predicted.

Now clearly we should not criticize Adams's and Leverrier's behaviour as 'unscientific'—after all, it led to the discovery of a new planet. But they did precisely what Popper criticized the Marxists for doing. They began with a theory—Newton's theory of gravity—which made an incorrect prediction about Uranus' orbit. Rather than concluding that Newton's theory must be wrong, they stuck by the theory and attempted to explain away the conflicting observations by postulating a new planet. Similarly, when capitalism showed no signs of giving way to communism, Marxists did not conclude that Marx's theory must be wrong, but stuck by the theory and tried to explain away the conflicting observations in other ways. So surely it is unfair to accuse Marxists of engaging in pseudo-science if we allow that what Adams and Leverrier did counted as good, indeed exemplary, science?

This suggests that Popper's attempt to demarcate science from pseudo-science cannot be quite right, despite its initial plausibility. For the Adams/Leverrier example is by no means atypical. In general, scientists do not just abandon their theories whenever they conflict with the observational data. Usually they look for ways of eliminating the conflict without having to give up their theory; see Chapter 5. Also, it is worth remembering that virtually every scientific theory conflicts with some observations—finding a theory that fits *all* the data perfectly is extremely difficult. Obviously if a theory persistently conflicts with more and more data, and no plausible way of explaining away the conflict is found, it will eventually have to be rejected. But little progress would be made if scientists simply abandoned their theories at the first sign of trouble.

The failure of Popper's demarcation criterion throws up an important question. Is it actually possible to find some common feature shared by all and only the things we call 'science'? Popper assumed that the answer was yes. He felt that Freud's and Marx's theories were clearly unscientific, so there must be some feature which they lack and which genuine scientific theories possess. But whether or not we accept Popper's negative assessment of Freud and Marx, his assumption that science has an 'essential nature' is questionable. After all, science is a heterogeneous activity, encompassing a wide range of disciplines and theories. It may be that they share some fixed set of features which define what it is to be a science, but it may not. The philosopher Ludwig Wittgenstein argued that there is no fixed set of features that define what it is to be a 'game'. Rather there is a loose cluster of features most of which are possessed by most games. But any particular game may lack any of the features in the cluster and still be a game. The same may be true of science. If so, a simple criterion for demarcating science from pseudo-science is unlikely to be found.

Chapter 2
Scientific inference

Scientists often tell us things about the world that we would not otherwise have believed. For example, biologists tell us that we are closely related to chimpanzees, geologists tell us that Africa and South America used to be joined together, and cosmologists tell us that the universe is expanding. But how did scientists reach these unlikely sounding conclusions? After all, no one has ever seen one species evolve from other, or a single continent split into two, or the universe getting bigger. The answer, of course, is that scientists arrived at these beliefs by a process of *reasoning* or *inference*. But it would be nice to know more about this process. What exactly is the nature of scientific inference?

Deduction and induction

Logicians make an important distinction between *deductive* and *inductive* inference, or deduction and induction for short. An example of a deductive inference is the following:

> All Frenchmen like red wine
> Pierre is a Frenchman
>
> ---
>
> Therefore, Pierre likes red wine

The two statements above the line are called the *premises* of the inference, while the statement below the line is called the *conclusion*. This is a deductive inference because it has the following property: *if* the premises are true, then the conclusion must be true too. If it's true that all Frenchmen like red wine, and that Pierre is a Frenchman, it follows that Pierre does indeed like red wine. This is sometimes expressed by saying that the premises of the inference *entail* the conclusion. Of course the premises of this inference are almost certainly not true—there are bound to be Frenchmen who do not like red wine. But that is not the point. What makes the inference deductive is the existence of an appropriate relation between premises and conclusion, namely that the truth of the premises guarantees the truth of the conclusion.

Not all inferences are deductive. Consider the following example:

> The first five eggs in the box were good.
> All the eggs have the same best-before date stamped on them.
>
> ---
>
> Therefore, the sixth egg will be good too.

This looks like a perfectly sensible piece of reasoning. But nonetheless it is not deductive, for the premises do not entail the conclusion. Even if the first five eggs were good, and all the eggs do have the same date stamp, it is quite conceivable that the sixth egg will be rotten. That is, it is logically possible for the premises of this inference to be true and yet the conclusion false, so the inference is not deductive. Instead it is known as an *inductive* inference. In a typical inductive inference, we move from premises about objects that we have examined to conclusions about objects of the same sort that we haven't examined—in this example, eggs.

Deductive inference is safer than its inductive cousin. When we reason deductively, we can be certain that if we start with true premises we will end up with a true conclusion. By contrast, inductive reasoning is quite capable of taking us from true premises to a false conclusion. Despite this defect, we seem to rely on inductive reasoning throughout our lives. For example, when you turn on your computer in the morning, you are confident it will not explode in your face. Why? Because you turn on your computer every morning, and it has never exploded up to now. But the inference from 'up until now, my computer has not exploded when I turned it on' to 'my computer will not explode this time' is inductive, not deductive. It is logically possible that your computer will explode this time, even though it has never done so before.

Do scientists use inductive reasoning too? The answer seems to be yes. Consider the condition known as Down's syndrome (DS). Geneticists tell us that people with DS have three copies of chromosome 21 instead of the usual two. How do they know this? The answer, of course, is that they examined a large number of people with DS and found that each had an additional copy of chromosome 21. They then reasoned inductively to the conclusion that *all* people with DS, including those they hadn't examined, have an additional copy. This inference is inductive not deductive. For it is possible, though unlikely, that the sample examined was unrepresentative. This example is not an isolated one. In effect, scientists reason inductively whenever they move from limited data to a more general conclusion, which they do all the time.

The central role of induction in science is sometimes obscured by how we talk. For example, you might read a newspaper report which says that scientists have found 'experimental proof' that genetically modified maize is safe to eat. What this means is that the scientists have tested the maize on a large number of people and none have come to any harm. But strictly speaking this doesn't *prove* that the maize is safe, in the sense in which mathematicians can prove Pythagoras' theorem, say. For the inference from 'the maize didn't harm any of the people on whom it was tested' to 'the maize will not harm anyone' is inductive, not deductive. The newspaper report should really have said that scientists have found good *evidence* that the maize is safe for humans. The word 'proof' should strictly only be used when we are dealing with deductive inferences. In this strict sense of the word, scientific hypotheses can rarely if ever be proved true by the data.

Most philosophers think it's obvious that science relies heavily on induction, indeed so obvious that it hardly needs arguing for. But remarkably, this was denied by the philosopher Karl Popper, whom we met in the last chapter. Popper claimed that scientists only need to use deductive inferences. This would be nice if it were true, for deductive inferences are safer than inductive ones, as we have seen.

Popper's basic argument was this. Although a scientific theory (or hypothesis) can never be proved true by a finite amount of data, it can be proved false, or refuted. Suppose a scientist is testing the hypothesis that all pieces of metal conduct electricity. Even if every piece of metal they examine conducts electricity, this doesn't prove that the hypothesis is true, for reasons that we've seen. But if the scientist finds even one piece of metal that fails to conduct electricity, this conclusively refutes the theory. For the inference from 'this piece of metal does not conduct electricity' to 'it is false that all pieces of metal conduct electricity' is a deductive inference—the premise entails the conclusion. So if a scientist were trying to refute their theory, rather than establish its truth, their goal could be accomplished without the use of induction.

The weakness of Popper's argument is obvious. For the goal of science is not solely to refute theories, but also to determine which theories are true (or probably true). When a scientist collects experimental data, their aim *might* be to show that a particular theory—their arch-rival's theory perhaps—is false. But much more likely, they are trying to convince people that their own theory is true. And in order to do that, they will have to resort to inductive reasoning of some sort. So Popper's attempt to show that science can get by without induction does not succeed.

Hume's problem

Although inductive reasoning is not logically watertight, it seems like a sensible way of forming beliefs about the world. Surely the fact that the sun has risen every day in the past gives us good reason to believe that it will rise tomorrow? If you came across someone who professed to be entirely agnostic about whether the sun will rise tomorrow or not, you would regard them as very strange indeed, if not irrational.

But what justifies this faith we place in induction? How should we go about persuading someone who refuses to reason inductively that they are wrong? The 18th-century Scottish philosopher David Hume (1711–76) gave a simple but radical answer to this question. He argued that the use of induction cannot be rationally justified at all. Hume admitted that we use induction all the time, in everyday life and in science, but insisted that this was a matter of brute animal habit. If challenged to provide a good reason for using induction, we can give no satisfactory answer, he thought.

How did Hume arrive at this startling conclusion? He began by noting that whenever we make inductive inferences, we seem to presuppose what he called the 'uniformity of nature'. To see what Hume meant by this, recall our examples. We had the inference from 'the first five eggs in the box were good' to 'the sixth egg will be good'; from 'the Down's syndrome patients examined had an extra chromosome' to 'all those with Down's syndrome have an extra chromosome'; and from 'my computer has never exploded until now' to 'my computer will not explode today'. In each case, our reasoning seems to depend on the assumption that objects we haven't examined will be similar, in relevant respects, to objects of the same sort that we have examined. That assumption is what Hume means by the uniformity of nature.

But how do we know that the uniformity assumption is true? Can we perhaps prove its truth somehow? No, says Hume, we cannot. For it is easy to *imagine* a world where nature is not uniform but changes its course randomly from day to day. In such a world, computers might sometimes explode for no reason, water might sometimes intoxicate us without warning, and billiard balls might sometimes stop dead on colliding. Since such a non-uniform world is conceivable, it follows that we cannot prove that the uniformity assumption is true. For if we could, then the non-uniform universe would be a logical impossibility.

Even if we cannot prove the uniformity assumption, we might nonetheless hope to find good empirical evidence for its truth. After all, the assumption has always held good up to now, so surely this is evidence that it is true? But this begs the question, says Hume! Grant that nature has behaved largely uniformly up to now. We cannot appeal to this fact to argue that nature will continue to be uniform, says Hume, because this assumes that what has happened in the past is a reliable guide to what will happen in the future—which *is* the uniformity of nature assumption. If we try to argue for the uniformity assumption on empirical grounds, we end up reasoning in a circle.

The force of Hume's point can be appreciated by imagining how you would persuade someone who doesn't trust inductive reasoning that they should. You might say: 'look, inductive reasoning has worked pretty well up until now. By using induction scientists have split the atom, landed on the moon, and invented lasers. Whereas people who haven't used induction have died nasty deaths. They have eaten arsenic believing it would nourish them, and jumped off tall buildings believing they would fly. Therefore it will clearly pay you to reason inductively.' But this wouldn't convince the doubter. For to argue that induction is trustworthy because it has worked well up to now is to reason inductively. Such an argument would carry no weight with someone who doesn't *already* trust induction. That is Hume's fundamental point.

This intriguing argument has exerted a powerful influence on the philosophy of science. (Popper's attempt to show that science need only use deduction was motivated by his belief that Hume had shown the unjustifiability of induction.) The influence of Hume's argument is not hard to understand. For normally we think of science as the very paradigm of rational enquiry. We place great faith in what scientists tell us about the world. But science relies on induction, and Hume's argument seems to show that induction cannot be rationally justified. If Hume is right, the foundations on which science is built do not look as solid as we might have hoped. This puzzling state of affairs is known as *Hume's problem of induction.*

Philosophers have responded to Hume's problem in literally dozens of ways; this is still an active area of research today. One response says that to seek a 'justification of induction', or to bemoan the lack of one, is ultimately incoherent. Peter Strawson, an Oxford philosopher from the 1950s, defended this view with the following analogy. If someone worried whether a particular action was legal, they could consult the lawbooks and see what they say. But suppose someone worried about whether the law itself was legal. This is an odd worry indeed. For the law is the standard against which the legality of other things is judged, and it makes little sense to enquire whether the standard itself is legal. The same applies to induction, Strawson argued. Induction is one of the standards we use to decide whether someone's beliefs about the world are justified. So it makes little sense to ask whether induction itself is justified.

Has Strawson really succeeded in defusing Hume's problem? Some philosophers say yes, others say no. But most agree that it is very hard to see how there *could* be a satisfactory justification of induction. (Frank Ramsey, a famous Cambridge philosopher, wrote in 1919 that to ask for a justification of induction was 'to cry for the moon'.) Whether this is something that should worry us, or shake our faith in science, is a difficult question that you should ponder for yourself.

Inference to the best explanation

The inductive inferences we've examined so far have all had essentially the same structure. In each case, the premise has had the form 'all examined *Fs* have been *G*', and the conclusion the form 'other *Fs* are also *G*'. In short, these inferences take us from examined to unexamined instances of a given kind.

Such inferences are widely used in everyday life and in science, as we have seen. However, there is another common type of non-deductive inference which doesn't fit this simple pattern. Consider the following example:

> The cheese in the larder has disappeared, apart from a few crumbs.
> Scratching noises were heard coming from the larder last night.
>
> ---
>
> Therefore, the cheese was eaten by a mouse.

It is obvious that this inference is non-deductive: the premises do not entail the conclusion. For the cheese could have been stolen by the maid, who cleverly left a few crumbs to make it look like the handiwork of a mouse; and the scratching noises could have been caused by the boiler overheating. Nonetheless, the inference is clearly a reasonable one. For the hypothesis that a mouse ate the cheese seems to provide a *better explanation* of the data than the 'maid and boiler' hypothesis. After all, maids do not normally steal cheese, and modern boilers rarely overheat. Whereas mice do eat cheese when they get the chance, and do make scratching sounds. So although we cannot be certain that the mouse hypothesis is true, on balance it looks plausible.

Reasoning of this sort is known as 'inference to the best explanation', or IBE for short. Certain terminological confusions surround the relation between IBE and induction. Some philosophers describe IBE as a *type* of inductive inference; in effect, they use 'inductive inference' to mean 'any inference which is not deductive'. Others *contrast* IBE with induction, as we have done. On this way of cutting the pie, 'induction' is reserved for inferences from examined to unexamined instances of a given kind; IBE and induction are then two different types of non-deductive inference. Nothing hangs on which choice of terminology we favour, so long as we stick to it consistently.

Scientists frequently use IBE. For example, Darwin argued for his theory of evolution by calling attention to various facts about the living world which are hard to explain if we assume that current species have been separately created, but which make perfect sense if current species have descended from common ancestors, as his theory held. For example, there are close anatomical similarities between the legs of horses and zebras. How do we explain this, if God created horses and zebras separately? Presumably he could have made their legs as different as he pleased. But if horses and zebras have descended from a common ancestor, this provides an obvious explanation of their anatomical similarity. Darwin argued that the ability of his theory to explain such facts constituted strong evidence for its truth. 'It can hardly be supposed', he wrote, 'that a false theory would explain, in so satisfactory a manner as does the theory of natural selection, the several large classes of fact above specified.'

Another example of IBE is Einstein's famous work on Brownian motion—the zig-zag motion of microscopic particles suspended in a liquid or gas. A number of attempted explanations of Brownian motion were advanced in the 19th century. One theory attributed the motion to electrical attraction between particles, another to agitation from external surroundings, and another to convection currents in the fluid. The correct explanation is based on the kinetic theory of matter, which says that liquids and gases are made up of atoms or molecules in motion. The suspended particles collide with the surrounding molecules, causing their erratic movements. This theory was proposed in the late 19th century but not widely accepted, not least because many scientists didn't believe that atoms and molecules were real entities. But in 1905, Einstein provided an ingenious mathematical treatment of Brownian motion, making a number of predictions that were later confirmed experimentally. After Einstein's work, the kinetic theory was quickly agreed to provide a better explanation of Brownian motion than the alternatives, and scepticism about the existence of atoms and molecules subsided.

The basic idea behind IBE—reasoning from one's data to a theory or hypothesis that explains the data—is straightforward. But how do we decide which of the competing hypotheses provides the 'best explanation' of the data? What criteria determine this? One popular answer is that a good explanation should be simple, or parsimonious. Consider again the cheese-in-the-larder example. There are two pieces of data that need explaining: the missing cheese and the scratching noises. The mouse hypothesis postulates just one cause—a mouse—to explain both pieces of data. But the maid-and-boiler hypothesis must postulate two causes—a dishonest maid and an overheating boiler—to explain the same data. So the mouse hypothesis is more parsimonious, hence better. The Darwin example is similar. Darwin's theory could explain a diverse range of facts about the living world, not just anatomical similarities between species. Each of these facts could in principle be explained in other ways, but the theory of evolution explained all the facts in one go—that is what made it the best explanation of the data.

The idea that simplicity or parsimony is the mark of a good explanation is quite appealing, and helps flesh out the abstract idea of IBE. But if scientists use simplicity as a guide to inference, this raises a deep question. Do we have reason to think that the universe is simple rather than complex? Preferring a theory which explains the data in terms of the fewest number of causes seems sensible. But are there any objective grounds for thinking that such a theory is more likely to be true than a less simple rival? Or is simplicity something that scientists value because it makes their theories easier to formulate and to understand? Philosophers of science do not agree on the answer to this difficult question.

Causal inference

A key goal of science is to discover the causes of natural phenomena. Often this quest is successful. For example, climate change scientists know that burning fossil fuels causes global warming; chemists know that heating a liquid causes it to become a gas; and epidemiologists know that the MMR vaccine does not cause autism. Since causal connections are not directly observable (as David Hume famously argued), scientific knowledge of this sort must be the result of inference. But how exactly does causal inference work?

It is helpful to distinguish two cases: inferring the cause of a particular event versus inferring a general causal principle. To illustrate the distinction, consider the contrast between 'a meteorite strike caused the extinction of the dinosaurs' and 'smoking causes lung cancer'. The former is a singular statement about the cause of a particular historical event, the latter a general statement about the cause of a certain *sort* of event (getting lung cancer). In both cases a process of inference has led scientists to believe the statements in question, but the inferences work in somewhat different ways. Here we focus on inferences of the second sort, i.e. to general causal principles.

Suppose a medical researcher wishes to test the hypothesis that obesity causes depression. How should they proceed? A natural first step is to see whether the two attributes are correlated. To assess this, they could examine a large sample of obese people, and see whether the incidence of depression is higher in this group than in the general population. If it is, then unless there is some reason to think the sample unrepresentative, it is reasonable to infer (by ordinary induction) that obesity and depression are correlated in the overall population.

Would such a correlation show that obesity causes depression? Not necessarily. First-year science students are routinely taught that correlation does not imply causation, and with good reason. For there are other possible explanations of the correlation. The direction of causation could be the other way round, i.e. being depressed might cause people to eat more, hence to become obese. Or there might be no causal influence of obesity on depression nor vice versa, but the two conditions are joint effects of a common cause. For example, perhaps low income raises the chance of obesity and also raises the chance of depression via a separate causal pathway (see Figure 3). If so, we would expect obesity and depression to be correlated in the population. This 'common cause' scenario is a major reason why causation cannot always be reliably inferred from correlational data.

How could we test the hypothesis that low income causes both obesity and depression? The obvious thing to do is to find a sample of individuals *all with the same income level*, and examine

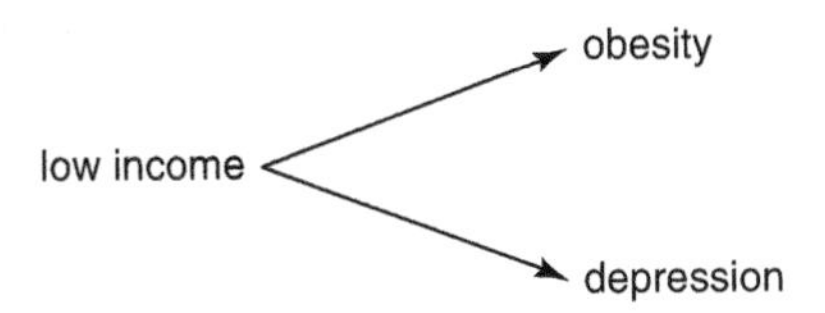

3. Causal graph depicting the hypothesis that low income is a common cause of both obesity and depression.

whether obesity and depression are correlated within the sample. If we do this for a number of different income levels, and find that within each income-homogeneous sample the correlation disappears, this is strong evidence in favour of the common cause hypothesis. For it shows that once income is taken into account, obesity is no longer associated with depression. Conversely, if a strong obesity–depression correlation exists even among individuals with the same income level, this is evidence against the common cause hypothesis. In statistical jargon, this procedure is known as 'controlling for' the variable income.

The underlying logic here is similar to that of the controlled experiment, a mainstay of modern science. Suppose an entomologist wishes to test the hypothesis that rearing insect larvae at higher temperatures leads to reduced adult body size. To test this, the entomologist gets a large number of insect larvae, rears some at a cool and others at a warm temperature, then measures the size of the resulting adults. For this to be an effective test of the causal hypothesis, it is important that all factors other than temperature be held constant between the two groups, so far as possible. For example, the larvae should all be from the same species, the same sex, and be fed the same food. So the entomologist must design their experiment carefully, controlling for all variables that could potentially affect adult body size. Only then can a difference in adult body size between the two groups safely be attributed to the temperature difference.

It is sometimes argued that controlled experiments are the only reliable way of making causal inferences in science. Proponents of this view argue that purely observational data, without any experimental intervention, cannot give us knowledge of causality. However this is a controversial thesis. For while controlled experimentation is certainly a good way of probing nature's secrets, the technique of statistical control can often accomplish something quite similar. In recent years, statisticians and computer scientists have developed powerful techniques for making causal inferences from observational data. Whether there is a fundamental methodological difference between experimental and observational data, vis-à-vis the reliability of the causal inferences that can be drawn from them, is a matter of continuing debate.

In modern biomedical science, a particular sort of controlled experiment is often given particular prominence. This is the *randomized controlled trial* (RCT), originally devised by R. A. Fisher in the 1930s, and often used to test the effectiveness of a new drug. In a typical RCT, patients with a particular medical condition, e.g. severe migraine, are divided into two groups. Those in the treatment group receive the drug, while those in the control group do not. The researchers then compare the two groups on the outcome of interest, e.g. relief of migraine symptoms. If those in the treatment group do significantly better than the control group, this is presumptive evidence that the drug works. The key feature of an RCT is that the initial division of the patients into two groups must be done at *random*. Fisher and his modern followers argue that this is necessary to sustain a valid causal inference.

Why is randomization so important? Because it helps to eliminate the effect of confounding factors on the outcome of interest. Typically the outcome will be affected by many factors, e.g. age, diet, and exercise. Unless all of these factors are known, the researcher cannot explicitly control for them. However by randomly allocating patients to the treatment and control groups, this problem can be largely circumvented. Even if factors other than the drug do affect the outcome, randomization ensures that any such factors are unlikely to be over-represented in either the treatment or the control group. So if there is a significant difference in outcome between treatment and control groups, this is very likely due to the drug. Of course this does not strictly *prove* that the drug was causally responsible, but it constitutes strong evidence.

In medicine, the RCT is usually regarded as the 'gold standard' for assessing causality. Indeed proponents of the movement known as 'evidence-based medicine' often argue that *only* an RCT can tell us when a particular treatment is causally effective. However this position is arguably too strong (and the appropriation of the word 'evidence' to refer only to RCTs is misleading). In many areas of science, RCTs are not feasible, for either practical or ethical reasons, and yet causal inferences are routinely made. Furthermore, much of the causal knowledge we have in everyday life we gained without RCTs. Young children know that putting their hand in the fire causes a painful burning sensation; no randomized trial was needed to establish this. While RCTs are certainly important, and should be done when feasible, it is not true that they are the only way of discovering causality.

Probability and scientific inference

Given that inductive reasoning cannot give us certainty, it is natural to hope that the concept of probability will help us understand how it works. Even if a scientist's evidence does not prove that their hypothesis is true, surely it can render it highly probable? Before exploring this idea we need to attend briefly to the concept of probability itself.

Probability has both an objective and a subjective guise. In its objective guise, probability refers to how often things in the world happen, or tend to happen. For example, if you are told that the probability of an Englishwoman living to age 90 is one in ten, you would understand this as meaning that one-tenth of all Englishwomen attain that age. Similarly, a natural understanding of the statement 'the probability that the coin will land heads is a half' is that in a long sequence of coin flips, the proportion of heads would be very close to a half. Understood this way, statements about probability are objectively true or false, independently of what anyone believes.

In its subjective guise, probability is a measure of rational degree of belief. Suppose a scientist tells you that the probability of finding life on Mars is extremely low. Does this mean that life is found on only a small proportion of all the celestial bodies? Surely not. For one thing, no one knows how many celestial bodies there are, nor how many of them contain life. So a different notion of probability is at work here. Now since there either is life on Mars or there isn't, talk of probability in this context must presumably reflect our ignorance of the state of the world, rather than describing an objective feature of the world itself. So it is natural to take the scientist's statement to mean that in the light of all the evidence, the rational degree of belief to have in the hypothesis that there is life on Mars is very low.

The idea that the rational degree of belief to have in a scientific hypothesis, given the evidence, may be viewed as a type of probability suggests a natural picture of how scientific inference works. Suppose a scientist is considering a particular hypothesis *H*. In the light of the evidence to date, the scientist has a certain degree of belief in *H*, denoted *P(H)*, which is a number between zero and one. (Another name for *P(H)* is the scientist's 'credence' in *H*.) Some new evidence then comes to light, e.g. from experiment or observation. In the light of this new evidence, the scientist updates their credence in *H* to $P_{new}(H)$. If the new evidence supports the theory, then $P_{new}(H)$ will be greater than *P(H)*, i.e. the scientist will have become more confident that *H* is true.

A toy example will help flesh this out. Suppose a playing card has been drawn from a well-shuffled pack and is concealed from your view. Let *H* be the hypothesis that the card is the queen of hearts. What is the value of *P(H)*, i.e. your initial rational credence in *H*? Presumably it is 1/52. For there are fifty-two cards in the pack and they are all equally likely to be chosen. Suppose you then learn that the chosen card is definitely a heart. Call this piece of information e. In the light of e, what is the value of $P_{new}(H)$, i.e. your updated credence in H given the new evidence? Clearly, $P_{new}(H)$ should equal 1/13—for there are thirteen hearts in the pack and you know that the concealed card is one of them. So learning e has increased your credence in H from 1/52 to 1/13.

This is all fairly obvious, but what is the general rule for updating your credence in the light of new information? The answer is called 'conditionalization'. To grasp this rule we need the concept of conditional probability. In the card example, $P(H)$ is your initial credence in hypothesis H. Your initial credence in H conditional on the assumption that e is true is denoted $P(H/e)$. (Read this as 'the probability of H given e'.) What is the value of $P(H/e)$? The answer is 1/13. For on the assumption that e is true, i.e. that the card drawn is a heart, your credence in the hypothesis H equals 1/13. When you learn that e *is* actually true, your new credence in H, i.e. $P_{new}(H)$, should then be set equal to your initial credence in H conditional on e, according to the rule of conditionalization.

Rule of conditionalization

Upon learning evidence e, $P_{new}(H)$ should equal $P(H/e)$.

To better understand the rule of conditionalization, note that the conditional probability $P(H/e)$ is by definition equal to the ratio $P(H \text{ and } e)/P(e)$. In the card example, $P(H \text{ and } e)$ denotes your initial credence that both H and e are true. But since in this case H logically entails e—for if the card is the queen of hearts then it must be a heart—it follows that $P(H \text{ and } e)$ is simply equal to $P(H)$, i.e. 1/52. What about $P(e)$? This is your initial credence that the chosen card is a heart. Since exactly one quarter of the cards in the deck are hearts, and you regard all the cards as equally likely to be the chosen one, it follows that $P(e)$ is ¼. Applying the definition of $P(H/e)$, this tells us that $P(H/e)$ equals 1/52 divided by ¼, which is 1/13—the same answer as we computed previously.

The rule of conditionalization may sound complicated, but like many logical rules we often obey it without thinking. In the card example, it is intuitively obvious that learning *e* should increase your rational credence in *H* from 1/52 to 1/13, and in practice this is exactly what most people would do. In doing so, they are implicitly obeying the rule of conditionalization even if they have never heard of it. In addition to its implicit uses, the conditionalization rule is often used explicitly by scientists, for example in certain sorts of statistical reasoning. The branch of statistics known as Bayesian statistics makes extensive use of updating by conditionalization. (The name 'Bayesian' refers to the 17th-century English clergyman Thomas Bayes, an early pioneer of probability theory, who discovered the conditionalization rule.)

Some philosophers of science wish to use updating by conditionalization as a general model for scientific inference, applicable even to inferences that are not explicitly probabilistic. The idea is that any rational scientist can be thought of as having an initial credence in their theory or hypothesis, which they then update in the light of new evidence by following the rule of conditionalization. Even if the scientist's conscious reasoning process looks nothing like this, it is a useful idealization according to these philosophers.

This 'Bayesian' view of scientific inference is quite attractive, as it sheds light on certain aspects of the scientific method. Consider the fact that when a scientific theory makes a testable prediction that turns out to be true, this is usually taken as evidence in favour of the theory. In Chapter 1 we had the example of Einstein's theory of general relativity predicting that starlight would be deflected by the sun's gravitational field; when this prediction was confirmed it increased scientists' confidence in Einstein's theory. But why should a successful prediction enhance a scientist's confidence in a theory, given that there will always be other possible explanations that can't be ruled out? Is this simply a brute fact about how scientists reason, or does it have a deeper explanation?

Bayesians argue that it does indeed have a deeper explanation. Suppose that a theory T entails a testable statement e. The scientist initially has credence $P(T)$ that T is true and $P(e)$ that e is true. We assume that both $P(T)$ and $P(e)$ take non-extreme values, i.e. are not zero or one. Suppose the scientist then learns that e is definitely true. If they follow the rule of conditionalization, their new credence in theory T, i.e. $P_{new}(T)$, must then be greater than $P(T)$ as a matter of logic. In other words, upon learning that their theory has made a true prediction, a scientist will necessarily increase their confidence in the theory so long as they obey the conditionalization rule. So the fact that successful predictions typically lead scientists to become more confident of their theories has a neat explanation, on the Bayesian view of scientific inference.

However the Bayesian view has its limitations. Much interesting scientific inference involves inventing theories or hypotheses that have never been thought of before. The great scientific advances made by Copernicus, Newton, and Darwin were all of this sort. Each of these scientists came up with a new theory which their predecessors had never entertained. The reasoning that led them to these theories cannot plausibly be regarded as Bayesian. For conditionalization describes how a scientist's rational credence in a theory should change when they get new evidence; this presumes that the theory has already been thought of. So scientific inferences that go from data to completely new theory cannot be understood in terms of conditionalization.

Another limitation of the Bayesian view concerns the source of the initial credences, prior to updating on the new evidence. In the card example, your initial rational credence that the chosen card was the queen of hearts was easy to determine, because there are fifty-two cards in a deck each with an equal chance of being chosen. But many scientific hypotheses are not like this. Consider the hypothesis that global warming will exceed four degrees by the year 2100. What should a scientist's initial credence in this hypothesis, before getting any relevant evidence, be? There is no obvious answer to this question. Some Bayesian philosophers of science reply that initial credences are purely subjective, i.e. they simply represent a scientist's 'best guess' about the hypothesis, so any initial credence is as good as any other. On this version of the Bayesian view, there is an objectively rational way for a scientist to *change* their credences when they get new evidence, i.e. conditionalization, but no objective constraint on what their initial credences should be.

This intrusion of a subjective dimension is regarded as unwelcome by many philosophers, leading them to conclude that the Bayesian view cannot be the whole story about scientific inference. Also, it shows that there cannot be a Bayesian 'solution' to Hume's problem of induction. The idea that we can somehow escape Hume's problem by invoking probability is an old one. Even if the sun's having risen every day in the past doesn't prove that it will rise tomorrow, surely it makes it highly probable? Whether this response to Hume ultimately works is a complex matter, but we can say the following. If the only objective constraints concern how we should change our credences, but what our initial credences should be is entirely subjective, then individuals with very bizarre opinions about the world will count as perfectly rational. So a probabilistic escape from Hume's problem will not fall out of the Bayesian view of scientific inference.

Chapter 3
Explanation in science

One important aim of science is to try and explain what happens in the world around us. Sometimes we seek explanations for practical ends. For example, we might want to know why the ozone layer is being depleted so quickly in order to try and do something about it. In other cases we seek scientific explanations simply to satisfy our intellectual curiosity—we want to understand more about how the world works. Historically, the pursuit of scientific explanation has been motivated by both goals.

Quite often, modern science is successful in its aim of supplying explanations. For example, chemists can explain why sodium turns yellow when it burns. Astronomers can explain why solar eclipses occur when they do. Economists can explain why the yen declined in value in the 1980s. Geneticists can explain why male baldness tends to run in families. Neurophysiologists can explain why extreme oxygen deprivation leads to brain damage. You can probably think of many other examples of successful scientific explanations.

But what exactly *is* a scientific explanation? What exactly does it mean to say that a phenomenon can be 'explained' by science? This is a question that has exercised philosophers since Aristotle, but our starting-point will be a famous account of scientific explanation put forward in the 1950s by the German-American philosopher

Carl Hempel. Hempel's account is known as the *covering law* model of explanation, for reasons that will become clear.

Hempel's covering law model of explanation

The basic idea behind the covering law model is straightforward. Hempel noted that scientific explanations are usually given in response to what he called 'explanation-seeking why-questions'. These are questions such as 'why is the earth not perfectly spherical?' or 'why do women live longer than men?'—they are demands for explanation. To give a scientific explanation is thus to provide a satisfactory answer to an explanation-seeking why-question. If we could determine the essential features that such an answer must have, we would know what scientific explanation is.

Hempel suggested that scientific explanations typically have the logical structure of an *argument*, i.e. a set of premises followed by a conclusion. The conclusion states that the phenomenon which needs explaining occurs, and the premises tell us why the conclusion is true. Thus suppose someone asks why sugar dissolves in water. This is an explanation-seeking why-question. To answer it, says Hempel, we must construct an argument whose conclusion is 'sugar dissolves in water' and whose premises tell us why this conclusion is true. The task of providing an account of scientific explanation then becomes the task of characterizing precisely the relation that must hold between a set of premises and a conclusion, in order for the former to count as an explanation of the latter. That was the problem Hempel set himself.

Hempel's answer to the problem was threefold. First, the premises should entail the conclusion, i.e. the argument should be a deductive one. Secondly, the premises should all be true. Thirdly, the premises should consist of at least one general law. General laws are things such as 'all metals conduct electricity', 'a body's acceleration varies inversely with its mass', and 'all plants contain chlorophyll'; they contrast with particular facts such as 'this piece of metal conducts electricity' and 'the plant on my desk contains chlorophyll'. General laws are sometimes called *laws of nature*. Hempel allowed that a scientific explanation could appeal to particular facts as well as general laws, but he held that at least one general law was always essential. So to explain a phenomenon, on Hempel's conception, is to show that its occurrence follows deductively from a general law, perhaps supplemented by other laws and/or particular facts, all of which must be true.

To illustrate, suppose I am trying to explain why the plant on my desk has died. I might offer the following explanation. Owing to the poor light in my study, no sunlight has been reaching the plant; but sunlight is necessary for a plant to photosynthesize; and without photosynthesis a plant cannot make the carbohydrates it needs to survive, and so will die; therefore my plant died. This explanation fits Hempel's model exactly. It explains the death of the plant by deducing it from two true laws—that sunlight is necessary for photosynthesis, and that photosynthesis is necessary for survival—and one particular fact—that the plant was not getting any sunlight. Given the truth of the two laws and the particular fact, the death of the plant *had* to occur; that is why the former constitute a good explanation of the latter.

Schematically, Hempel's model of explanation can be written as follows:

General Laws
Particular Facts
⇒
Phenomenon to be explained

The phenomenon to be explained is called the *explanandum*, and the general laws and particular facts that do the explaining are called the *explanans*. The *explanandum* may be either particular or general. In the previous example, it was a particular fact—the death of my plant. But sometimes the things we want to explain are general. For example, we might wish to explain why exposure to the sun often leads to skin cancer. This is itself a generality, not a particular fact. To explain it, we would need to deduce it from more fundamental laws—presumably, laws about the impact of radiation on skin cells, combined with particular facts about the amount of radiation in sunlight. So the structure of a scientific explanation is essentially the same whether the *explanandum*, i.e. thing we are trying to explain, is particular or general.

It is easy to see where the covering law model gets its name. For according to the model, the essence of explanation is to show that the phenomenon to be explained is 'covered' by some general law of nature. There is certainly something appealing about this idea. For showing that a phenomenon is a consequence of a general law takes the mystery out of it—it renders it more intelligible. And many actual scientific explanations do fit the pattern Hempel describes. For example, Newton explained why the planets move in ellipses around the sun by showing that this can be deduced from his law of universal gravitation, along with some minor additional assumptions. Newton's explanation fits Hempel's model exactly: a phenomenon is explained by showing that it had to be so, given the laws of nature plus some additional facts. After Newton, there was no longer any mystery about why planetary orbits are elliptical.

Hempel was aware that not all scientific explanations fit his model exactly. For example, if you ask someone why the smog in Athens has worsened in recent years they might reply 'because of the increase in domestic wood-burning'. This is true, and is a perfectly acceptable scientific explanation, though it involves no mention of any laws. But Hempel would say that if the explanation were spelled out in full detail, laws would enter the picture. Presumably there is a law which says something like 'if wood-smoke emissions exceed a certain level in an area of a given size, and if the wind is sufficiently light, smog clouds will form'. The full explanation of why the smog in Athens has worsened would cite this law, along with the fact that wood-burning in Athens has increased and that wind levels there are fairly low. In practice we wouldn't spell out the explanation in this much detail unless we were being very pedantic. But if we were to spell it out, it would correspond quite well to the covering law pattern.

Hempel drew an interesting consequence from his model about the relation between explanation and prediction. He argued that these are two sides of the same coin. Whenever we give a covering law explanation of a phenomenon, the laws and particular facts we cite would have enabled us to predict the occurrence of the phenomenon, if we hadn't already known about it. To illustrate, consider again Newton's explanation of why planetary orbits are elliptical. This fact was known long before Newton explained it using his theory of gravity—it was discovered by Kepler. But had it not been known, Newton would have been able to predict it from his theory of gravity. Hempel expressed this by saying that every scientific explanation is potentially a prediction—it would have served to predict the phenomenon in question, had it not already been known. The converse is also true, Hempel thought: every reliable prediction is potentially an explanation. To illustrate, suppose scientists predict that mountain gorillas will be extinct by 2030, based on information about the destruction of their habitat. Suppose they turn out to be right. According to Hempel, the information they used to predict the gorillas' extinction before it happened will serve to explain that same fact after it has happened. Explanation and prediction are structurally symmetric.

Though the covering law model captures the structure of many actual scientific explanations quite well, it also faces a number of awkward counterexamples. In particular, there are cases that fit the covering law model but intuitively do not count as genuine scientific explanations. These cases suggest that Hempel's model is too liberal—it allows in things that should be excluded. We focus on two such cases here.

Case (i): the problem of symmetry

Suppose you are lying on the beach on a sunny day, and you notice that a flagpole is casting a shadow of 20 metres across the sand (see Figure 4).

Someone asks you to explain why the shadow is 20 metres long. This is an explanation-seeking why-question. A plausible answer might go as follows: 'light rays from the sun are hitting the flagpole, which is exactly 15 metres high. The angle of elevation of the sun is 37°. Since light travels in straight lines, a simple trigonometric calculation (tan 37° = 15/20) shows that the flagpole will cast a shadow 20 metres long.'

This looks like a perfectly good scientific explanation. And by rewriting it in accordance with Hempel's schema, we can see that it fits the covering law model:

General laws	Light travels in straight lines Laws of trigonometry
Particular facts	Angle of elevation of sun is 37° Flagpole is 15 metres high
Phenomenon to be explained	Shadow is 20 metres long

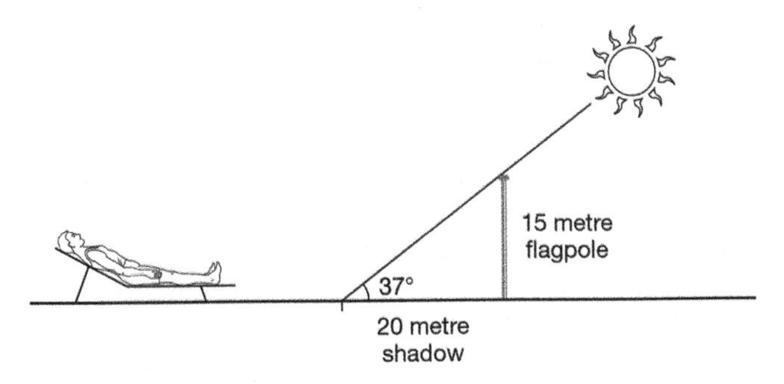

4. A 15-metre flagpole casts a shadow of 20 metres when the sun is 37° overhead.

The length of the shadow is deduced from the height of the flagpole and the angle of elevation of the sun, along with the optical law that light travels in straight lines and the laws of trigonometry. Since these laws are true, and since the flagpole is indeed 15 metres high, the explanation satisfies Hempel's requirements precisely. So far so good. The problem arises as follows. Suppose we swap the *explanandum*—that the shadow is 20 metres long—with the particular fact that the flagpole is 15 metres high. The result is this:

General laws	Light travels in straight lines
	Laws of trigonometry
Particular facts	Angle of elevation of sun is 37°
	Shadow is 20 metres long
Phenomenon to be explained	Flagpole is 15 metres high

This 'explanation' clearly conforms to the covering law pattern too. The height of the flagpole is deduced from the length of the shadow it casts and the angle of elevation of the sun, along with the optical law that light travels in straight lines and the laws of trigonometry. But it seems very odd to regard this as an *explanation* of why the flagpole is 15 metres high. The real explanation of why the flagpole is 15 metres high is presumably that a carpenter deliberately made it so—it has nothing to do with the length of the shadow that it casts. So Hempel's model is too liberal: it allows something to count as a scientific explanation which obviously is not.

The general moral of the flagpole example is that the concept of explanation exhibits an important asymmetry. The height of the flagpole explains the length of the shadow, given the relevant laws and additional facts, but not vice versa. In general, if x explains y, given the relevant laws and additional facts, then it will not be true that y explains x, given the same laws and facts. This is sometimes expressed by saying that explanation is an asymmetric relation. Hempel's covering law model does not respect this asymmetry. For just as we can deduce the length of the shadow from the height of the flagpole, given the laws and additional facts, so we can deduce the height of the flagpole from the length of the shadow. So Hempel's model fails to capture fully what it is to be a scientific explanation, for it implies that explanation should be a symmetric relation when in fact it is asymmetric.

The shadow and flagpole case also provides a counterexample to Hempel's thesis that explanation and prediction are two sides of the same coin. The reason is obvious. Suppose you didn't know how high the flagpole was. If someone told you that it was casting a shadow of 20 metres and that the sun was 37° overhead, you would be able to *predict* the flagpole's height, given that you knew the relevant optical and trigonometrical laws. But as we have just seen, this information clearly doesn't *explain* why the flagpole has the height it does. So in this example prediction and explanation part ways. Information that serves to predict a fact before we know it does not serve to explain that same fact after we know it, which contradicts Hempel's thesis.

Case (ii): the problem of irrelevance

Suppose a young child is in a maternity ward in a hospital. The child notices that one person in the room—who is a man called John—is not pregnant, and asks the doctor why not. The doctor replies: 'John has been taking birth control pills regularly for the last few years. People who take birth control pills regularly never become pregnant. Therefore, John has not become pregnant.' Let us suppose that what the doctor says is true—John is mentally ill and does indeed take birth control pills, which he believes help him. Even so, the doctor's reply to the child is clearly not helpful. The correct explanation of why John has not become pregnant, obviously, is that he is male and males cannot become pregnant.

However, the explanation the doctor has given fits the covering law model exactly. The doctor deduces the phenomenon to be explained—that John is not pregnant—from the general law that people who take birth control pills do not become pregnant and the particular fact that John has been taking birth control pills. Since both the general law and the particular fact are true, and since they do entail the *explanandum*, according to the covering law model the doctor has given an explanation of why John is not pregnant. But of course he hasn't.

The general moral is that a good explanation of a phenomenon should contain information that is *relevant* to the phenomenon's occurrence. This is where the doctor's reply to the child goes wrong. Although what the doctor tells the child is perfectly true, the fact that John has been taking birth control pills is irrelevant to his not being pregnant, because he wouldn't have been pregnant even if he hadn't been taking the pills. This is why the doctor's reply does not constitute a good answer to the child's question. Hempel's model does not respect this crucial feature of our concept of explanation.

Explanation and causality

Since the covering law model encounters problems, it is natural to look for an alternative way of understanding scientific explanation. Some philosophers believe that the key lies in the concept of *causality*. This is quite an attractive suggestion. For in many cases to explain a phenomenon is indeed to say what caused it. For example, if an accident investigator is trying to explain an aeroplane crash, they are obviously looking for the cause of the crash. Indeed the questions 'why did the plane crash?' and 'what was the cause of the plane crash?' are practically synonymous. Similarly, if an ecologist is trying to explain why there is less biodiversity in the tropical rainforests than there used to be, they are looking for the cause of the reduction in biodiversity. The link between the concepts of explanation and causality is quite intimate.

Impressed by this link, a number of philosophers have abandoned the covering law account of explanation in favour of causality-based accounts. The details vary, but the basic idea behind these accounts is that to explain a phenomenon is simply to say what caused it. In some cases, the difference between the covering law and causal accounts is not actually very great, for to deduce the occurrence of a phenomenon from a general law often just *is* to give its cause. For example, recall again Newton's explanation of why planetary orbits are elliptical. We saw that this explanation fits the covering law model—for Newton deduced the shape of the planetary orbits from his law of gravity, plus some additional facts. But Newton's explanation was also a causal one, since elliptical planetary orbits are *caused* by the gravitational attraction between planets and the sun.

However the covering law and causal accounts are not fully equivalent—in some cases they diverge. Indeed, many philosophers favour a causal account of explanation precisely because they think it can avoid some of the problems facing the covering law model. Recall the flagpole problem. Why do our intuitions tell us that the height of the flagpole explains the length of the shadow, given the laws, but not vice versa? Plausibly, because the height of the flagpole is the *cause* of the shadow being 20 metres long, but the shadow being 20 metres long is not the cause of the flagpole being 15 metres high. So unlike the covering law model, a causal account of explanation gives the 'right' answer in the flagpole case—it respects our intuition that we cannot explain the height of the flagpole by pointing to the length of the shadow it casts.

The general moral of the flagpole problem was that the covering law model cannot accommodate the fact that explanation is an asymmetric relation. Now causality is obviously an asymmetric relation too: if x is the cause of y, then y is not the cause of x. For example, if the short-circuit caused the fire, then the fire clearly did not cause the short-circuit. It is therefore natural to suggest that the asymmetry of explanation derives from the asymmetry of causality. If to explain a phenomenon is to say what caused it, then since causality is asymmetric we should expect explanation to be asymmetric too—as it is. The covering law model runs up against the flagpole problem precisely because it tries to analyse the concept of scientific explanation without reference to causality.

The same is true of the birth control pill case. That John takes birth control pills does not explain why he isn't pregnant, because the birth control pills are not the *cause* of his not being pregnant. Rather, John's sex is the cause of his not being pregnant. That is why we think that the correct answer to the question 'why is John not pregnant?' is 'because he is male, and males can't become pregnant', rather than the doctor's answer. So the covering law model runs into the problem of irrelevance precisely because it does not explicitly require that a scientific explanation identify the cause of the phenomenon that we wish to explain.

It is easy to criticize Hempel for failing to respect the close link between causality and explanation, as many philosophers have done. In some ways this criticism is a bit unfair. For Hempel subscribed to the philosophical doctrine called *empiricism*, and empiricists are traditionally suspicious of the concept of causality. Empiricism says that all our knowledge comes from experience. David Hume, whom we met in Chapter 2, was a leading empiricist, and he argued that it is impossible to experience causal relations. So he concluded that they don't exist—causality is something that we humans 'project' onto the world! This is a very hard conclusion to accept. Surely it is an objective fact that dropping glass vases causes them to break? Hume denied this. He allowed that it is an objective fact that most glass vases which have been dropped have in fact broken. But our idea of causality includes more than this. It includes the idea of a causal connection between the dropping and the breaking, i.e. that the former *brings about* the latter. No such connections are to be found in the world, according to

Hume: all we see is a vase being dropped, and then it breaking a moment later. This leads us to believe there is a causal connection between the two, but in reality there is not.

Few empiricists have accepted this startling conclusion outright. But as a result of Hume's work, they have tended to regard causality as a concept to be treated with caution. So to an empiricist, the idea of analysing explanation in terms of causality would seem perverse. If one's goal is to clarify the concept of scientific explanation, as Hempel's was, there is little point in using notions which are equally in need of clarification themselves. So the fact that the covering law model makes no mention of causality was not a mere oversight on Hempel's part. In recent years empiricism has declined somewhat in popularity. Furthermore, many philosophers have come to the conclusion that the concept of causality, although problematic, is indispensable to how we understand the world. So the idea of a causality-based account of scientific explanation seems more acceptable than it would have done in Hempel's day.

Causality-based accounts capture the structure of many actual scientific explanations quite well, but there are also cases they fit less well. Consider what are called 'theoretical identifications' in science, such as 'water is H_2O' or 'temperature is mean molecular kinetic energy'. In both cases, a familiar everyday concept is equated or identified with a more esoteric scientific concept. Such theoretical identifications furnish us with what appear to be scientific explanations. When chemists discovered that water is H_2O, they thereby explained what water is. Similarly, when physicists discovered that an object's temperature is the average kinetic energy of its molecules, they thereby explained what temperature is. But neither of these explanations is causal. Being made of H_2O doesn't *cause* a substance to be water—it just *is* being water. Having a particular mean molecular kinetic energy doesn't *cause* a liquid to have the temperature it does—it just *is* having that temperature. If these examples are accepted as legitimate scientific explanations, they suggest that causality-based accounts of explanation cannot be the whole story.

Can science explain everything?

Modern science can explain a great deal about the world we live in. But there are also numerous facts that have not been explained by science, or at least not explained fully. The origin of life is one such example. We know that about four billion years ago, molecules with the ability to make copies of themselves appeared in the primeval soup, and life evolved from there. But we do not understand how these self-replicating molecules got there in the first place (though some possible scenarios have been sketched). Another example is the fact that children with Asperger's syndrome often have very good memories. Numerous studies have confirmed this fact, but as yet nobody has succeeded in explaining it.

Many people believe that in the end, science will be able to explain facts of this sort. This is quite a plausible view. Molecular biologists are working hard on the problem of the origin of life, and only a pessimist would say they will never solve it. Admittedly the problem is not easy, not least because it is hard to know what conditions on earth four billion years ago were like. But nonetheless, there is no reason to think that the origin of life will never be explained. Similarly for the exceptional memories of children with Asperger's. The science of memory is still fairly new, and much remains to be discovered about the neurological basis of conditions such as Asperger's syndrome. Obviously we cannot guarantee that the explanation will eventually be found. But given the number of explanatory successes that modern science has already notched up, the smart money must be on many of today's unexplained facts eventually being explained too.

But does this mean that science can in principle explain everything? Or are there some phenomena that must forever elude scientific explanation? This is not an easy question to answer. On the one hand, it seems arrogant to assert that science can explain everything. On the other hand, it seems short-sighted to assert that any particular phenomenon can never be explained scientifically. For science changes and develops fast, and a phenomenon that looks completely inexplicable from the vantage-point of today's science may be easily explained tomorrow.

According to many philosophers, there is a purely logical reason why science will never be able to explain everything. For in order to explain something, whatever it is, we need to invoke something else. But what explains the second thing? To illustrate, recall that Newton explained a diverse range of phenomena using his law of gravity. But what explains the law of gravity itself? If someone asks *why* all bodies exert a gravitational attraction on each other, what should we tell them? Newton had no answer to this question. In Newtonian science the law of gravity was a fundamental principle: it explained other things, but could not itself be explained. The moral generalizes. However much the science of the future can explain, the explanations it gives will have to make use of certain fundamental laws and principles. Since nothing can explain itself, it follows that at least some of these laws and principles will themselves remain unexplained.

Whatever one makes of this argument, it is undeniably very abstract. It purports to show that some things will never be explained, but does not tell us what they are. However, some philosophers have made concrete suggestions about phenomena which they think science can never explain. An example is *consciousness*—the distinguishing feature of thinking, feeling creatures such as ourselves and other higher animals. Much research into the nature of consciousness has been and continues to be done, by neuroscientists, psychologists, and others. But a number of recent philosophers claim that whatever this research throws up, it will never fully explain the nature of consciousness. There is something intrinsically mysterious about the phenomenon of consciousness, they maintain, that no amount of scientific investigation can eliminate.

What are the grounds for this view? The basic argument is that conscious experiences are fundamentally unlike anything else in the world, in that they have a 'subjective aspect'. Consider for example the experience of watching a terrifying horror movie. This is an experience with a very distinctive 'feel' to it; in the current jargon, there is 'something that it is like' to have the experience. Neuroscientists may one day be able to give a detailed account of the complex goings-on in the brain which produce our feeling of terror. But will this explain why watching a horror movie feels the way it does, rather than feeling some other way? Some philosophers argue that it will not. On their view, the scientific study of the brain can at most tell us which brain processes are correlated with which conscious experiences. This is certainly interesting and valuable information. However it doesn't tell us *why* experiences with distinctive subjective 'feels' should result from the purely physical goings-on in the brain. Hence consciousness, or at least one important aspect of it, is scientifically inexplicable.

Though quite compelling, this argument is controversial and not endorsed by all philosophers, let alone all neuroscientists. Indeed a well-known 1991 book by the philosopher Daniel Dennett is defiantly entitled *Consciousness Explained*. Supporters of the view that consciousness is scientifically inexplicable are sometimes accused of having a lack of imagination. Even if it is true that brain science as currently practised cannot explain the subjective aspect of conscious experience, can we not imagine the emergence of a different type of brain science, with different explanatory techniques, that *does* explain why our experiences feel the way they do? There is a long tradition of philosophers trying to tell scientists what is and isn't possible, and later scientific developments have often proved the philosophers wrong. Only time will tell whether the same fate awaits those who argue that consciousness must always elude scientific explanation.

Explanation and reduction

The different scientific disciplines are designed for explaining different types of phenomena. To explain why rubber doesn't conduct electricity is a task for physics. To explain why turtles have such long lives is a task for biology. To explain why higher interest rates reduce inflation is a task for economics, and so on. In short, there is a division of labour between the different sciences: each specializes in explaining its own particular set of phenomena. This explains why the sciences are not usually in competition with one another—why biologists, for example, do not worry that physicists and economists might encroach on their turf.

Nonetheless, it is widely held that the different branches of science are not all on a par: some are more fundamental than others. Physics is usually regarded as the most fundamental science of all. Why? Because the objects studied by the other sciences are ultimately composed of physical particles. Consider living organisms, for example. Living organisms are made up of cells, which are themselves made up of water, nucleic acids, proteins, sugars, and lipids, all of which consist of molecules or long chains of molecules joined together. But molecules are made up of atoms, which are physical particles. So the objects biologists study are ultimately just very complex physical entities. The same applies to the other sciences, even the social sciences. Take economics for example. Economics studies the behaviour of firms and consumers in the market place, and the consequences of this behaviour. But consumers are human beings and firms are made up of human beings; and human beings are living organisms, hence physical entities.

Does this mean that, in principle, physics can subsume all the higher-level sciences? Since everything is made up of physical particles, surely if we had a complete physics, which allowed us to predict perfectly the behaviour of every physical particle in the universe, all the other sciences would become superfluous? Most philosophers resist this line of thought. After all, it seems crazy to suggest that physics might one day be able to explain the things that biology and economics explain. The prospect of deducing the laws of biology and economics straight from the laws of physics looks very remote. Whatever the physics of the future looks like, it is most unlikely to be capable of predicting economic downturns. Far from being reducible to physics, sciences such as biology and economics seem largely autonomous of it.

This leads to a philosophical puzzle. How can a science which studies entities which are ultimately physical *not* be reducible to physics? Granted that the higher-level sciences are in fact autonomous of physics, how is this possible? According to some philosophers, the answer lies in the fact that the objects studied by the higher-level sciences are *multiply realized* at the physical level. To illustrate the idea of multiple realization, imagine a collection of ashtrays. Each individual ashtray is obviously a physical entity, like everything else in the universe. But the physical composition of the ashtrays could be very different—some might be made of glass, others of aluminium, others of plastic, and so on. And they will probably differ in size, shape, and weight. There is virtually no limit on the range of different physical properties that an ashtray can have. So it is impossible to define the concept 'ashtray' in purely physical terms. We cannot find a true statement of the form 'x is an ashtray if and only if x is...' where the blank is filled by an expression taken from the language of physics. This means that ashtrays are multiply realized at the physical level.

Philosophers have often invoked multiple realization to explain why psychology cannot be reduced to physics or chemistry, but in principle the explanation works for any higher-level science. For example, consider the biological fact that nerve cells live longer than skin cells. Cells are physical entities, so one might think that this fact will one day be explained by physics. However, cells are almost certainly multiply realized at the microphysical level. Cells are ultimately made up of atoms, but the precise arrangement of atoms will be very different in different cells. So the concept 'cell' cannot be defined in terms drawn from fundamental physics. There is no true statement of the form 'x is a cell if and only if x is...' where the blank is filled by an expression taken from the language of microphysics. If this is correct, it means that fundamental physics will never be able to explain why nerve cells live longer than skin cells, or indeed any other facts about cells. The vocabulary of cell biology and the vocabulary of physics do not map onto each other in the required way. Thus we have an explanation of why it is that cell biology cannot be reduced to physics, despite the fact that cells are physical entities. Not all philosophers are happy with the doctrine of multiple realization, but it does promise a neat explanation of the autonomy of the higher-level sciences, both from physics and from each other.

Chapter 4
Realism and anti-realism

There is an ancient debate in philosophy between two opposing schools of thought called *realism* and *idealism*. Realism holds that the physical world exists independently of human thought and perception. Idealism denies this—it claims that the physical world is in some way dependent on the conscious activity of humans. To most people, realism seems more plausible than idealism. For realism fits well with the commonsense view that the facts about the world are 'out there' waiting to be discovered. Indeed at first glance idealism can sound plain silly. Since rocks and trees would continue to exist even if the human race died out, in what sense is their existence dependent on human minds? In fact the issue is a bit more subtle than this, and continues to be discussed by philosophers today.

Though the traditional realism/idealism issue belongs to an area of philosophy called *metaphysics*, it has nothing in particular to do with science. Our concern in this chapter is with a contemporary debate that is specifically about science, and is in some ways analogous to the traditional issue. The debate is between a position known as *scientific realism* and its converse, known as *anti-realism* or *instrumentalism*. From now on we shall use the word 'realism' to mean scientific realism, and 'realist' to mean scientific realist.

Scientific realism and anti-realism

The basic idea of scientific realism is straightforward. Realists hold that science aims to provide a true description of the world, and that it often succeeds. So a good scientific theory, according to realists, is one that truly describes the way the world is. This may sound like a fairly innocuous doctrine. For surely no one thinks that science is aiming to produce a false description of the world? But that is not what anti-realists think. Rather, anti-realists hold that the aim of science is to find theories that are *empirically adequate*, i.e. which correctly predict the results of experiment and observation. If a theory achieves perfect empirical adequacy, the further question of whether it truly describes the world is redundant, for anti-realists; indeed some argue that this question does not even make sense.

The contrast between realism and anti-realism is starkest for sciences which make claims about the unobservable region of reality. Physics is the obvious example. Physicists advance theories about atoms, electrons, quarks, leptons, and other strange entities, none of which can be observed in the normal sense of the word; moreover these theories are typically couched in a highly mathematical language. So physical theories are rather different from the commonsense descriptions of the world that non-scientists give. Nonetheless, realists argue, these theories *are* attempts to describe the world—the subatomic world—and the measure of their success is whether what they say about the world is true. In this respect, scientific theories and commonsense descriptions of the world are on a par.

Anti-realists argue that empirical adequacy, not truth, is the real aim of scientific theorizing. Physicists may talk about unobservable entities, but they are merely *convenient fictions* introduced in order to help predict observable phenomena. To illustrate, consider again the kinetic theory of gases, which says that any volume of gas contains a large number of very small entities in motion. These entities—molecules—are unobservable. From the kinetic theory we can deduce various consequences about the observable behaviour of gases, for example that heating a sample of gas will cause it to expand if the pressure remains constant, which can be verified experimentally. Anti-realists argue that the only purpose of positing unobservable entities in the kinetic theory is to deduce consequences of this sort. Whether or not gases really do contain molecules in motion doesn't matter; the point of the kinetic theory is not to truly describe the hidden facts, but just to provide a convenient way of predicting observations. We can see why anti-realism is sometimes called 'instrumentalism'—it regards scientific theories as instruments for helping us predict observable phenomena, rather than as attempts to describe the underlying nature of reality.

Since the realism/anti-realism debate concerns the aim of science, one might think it could be resolved by simply asking the scientists themselves. Why not do a straw poll of scientists asking them about their aims? But this suggestion misses the point—it takes the expression 'the aim of science' too literally. When we ask what the aim of science is, we are not asking about the aims of individual scientists. Rather, we are asking how best to make sense of what scientists say and do—how to *interpret* the scientific enterprise. While it would certainly be interesting to discover scientists' own views on the realism/anti-realism debate, the issue is ultimately a philosophical one.

One motivation for anti-realism stems from the belief that we cannot actually attain knowledge of the unobservable part of reality—it lies beyond human ken. This pessimistic belief stems from empiricism, the philosophical doctrine according to which human knowledge is limited to what can in principle be experienced. Applied to science, the empiricist doctrine becomes the view that the limits to scientific knowledge are set by our powers of observation. So science can give us knowledge of fossils, trees, and sugar crystals, but not of atoms, electrons, and quarks. This view is not altogether implausible. For no one could seriously doubt the existence of fossils and trees, but the same is not true of atoms and electrons. As we saw in the last chapter, in the late 19th century many leading scientists did doubt the existence of atoms. Anyone who accepts such a view must obviously give some explanation of *why* scientists advance theories that posit unobservable entities, if scientific knowledge is limited to what can be observed. The explanation anti-realists give is that they are convenient fictions, designed to help predict the behaviour of things in the observable world.

Realists do not agree that scientific knowledge is limited by our powers of observation. On the contrary, they believe that we already have substantial knowledge of unobservable reality. For there is every reason to believe that our best scientific theories are true, and those theories talk about unobservable entities. Consider for example the atomic theory of matter, which says that all matter is made up of atoms. The atomic theory is capable of explaining a great range of facts about the world. Realists regard this as good evidence that the theory is true, i.e. that matter really is made up of atoms which behave as the theory says. Of course the theory *might* be false, despite the apparent evidence in its favour, but so might any theory. Just because atoms are unobservable, that is no reason to interpret atomic theory as anything other than an attempted description of reality—and a very successful one, in all likelihood.

A different motivation for anti-realism stems from the fact that scientific theories have certain peculiarities which ordinary descriptions of the world do not. Much scientific theorizing involves the construction of models, often couched in mathematical language. Such models typically make idealizing assumptions that are known to be false of the real world, but are necessary to keep the model tractable. In economics, for example, many models assume that agents are perfectly rational, have perfect information, and make decisions that maximize their utility. Economists know that real people are not like this, but they hope that their models may nonetheless shed light on the real-world economy. Similarly, in evolutionary biology many models assume that the population size is infinite and that mating is random; these assumptions greatly simplify the mathematics. No real population satisfies these assumptions, but biologists hope that they are a good enough approximation to reality for their models to have explanatory power. Anti-realists often argue that the prevalence of idealized models in science supports their view. It makes no sense to regard such models as attempts to truly describe the world, they argue, since they contain assumptions known to be false. The aim of such models is empirical adequacy, not truth.

Realists do not regard this argument as decisive. The role of idealized models in scientific theorizing does not compel us to reject outright the idea that science aims at truth. Instead we need to accept that approximate truth, rather than exact truth, is the goal of such models, realists argue. Consider for example a mathematical model of climate change. Such a model will incorporate many simplifying assumptions that are known not to be exactly true, e.g. that fossil fuels are the sole source of carbon dioxide. But that does not mean that the model aims *only* at generating correct predictions. Rather, the model is aiming to provide an approximately true description of the hidden causal factors that actually affect climate change. Certainly a good climate change model should be predictively successful, but the real aim is to devise a model that accurately represents, as far as possible, the real causal influences on the climate. An idealized model will never be a literally true description of the world, but it may still be a good approximation, realists argue.

The 'no miracles' argument

Many theories which posit unobservable entities are empirically successful—they make excellent predictions about the behaviour of macroscopic objects. The kinetic theory of gases, described in the section 'Scientific realism and anti-realism', is one example, and there are many others. Furthermore, such theories often have technological applications. For example, laser technology is based on a theory about what happens when electrons in an atom go from higher to lower energy states. And lasers work—they allow us to correct myopia, print high-quality text, attack our enemies with guided missiles, and more. The theory that underpins laser technology is therefore highly empirically successful.

The empirical success of theories which posit unobservable entities is the basis of one of the main arguments for scientific realism, known as the 'no miracles' argument. Originally formulated by Hilary Putnam, a leading American philosopher, the no miracles argument says that it would be an extraordinary coincidence if a theory which posits electrons and atoms made accurate predictions unless these entities actually exist. If there are no atoms and electrons, what explains the theory's close fit with the empirical data? Similarly, how do we explain the technological advances our theories have led to, unless by supposing that the theories in question are true? If atoms and electrons are just 'convenient fictions', as anti-realists maintain, then why do lasers work? On this view, being an anti-realist is akin to believing in miracles. Since it is better not to believe in miracles if a non-miraculous alternative is available, we should be scientific realists.

This argument is not intended to *prove* that realism is right and anti-realism wrong. Rather it is a plausibility argument—an inference to the best explanation. The phenomenon to be explained is the fact that many theories which postulate unobservable entities and processes enjoy a high level of empirical success. The best explanation of this fact, say advocates of the no miracles argument, is that the theories are true—the entities in question really exist, and behave just as the theories say.

Unless we accept this explanation, the empirical success of our theories is an unexplained mystery.

One anti-realist response to the no miracles argument appeals to the history of science. Historically, there are many examples of scientific theories which were empirically successful in their day but later turned out to be false. In a well-known article from the 1980s, the American philosopher of science Larry Laudan listed more than thirty such theories, drawn from a range of different scientific disciplines and eras. The phlogiston theory of combustion is one example. This theory, which was widely accepted until the end of the 18th century, held that when any object burns it releases a substance called 'phlogiston' into the atmosphere. Modern chemistry teaches us that this is false: there is no such substance as phlogiston. Rather, burning occurs when things react with oxygen in the air. But despite the non-existence of phlogiston, the phlogiston theory was empirically quite successful: it fitted the data available at the time reasonably well.

Examples such as this suggest that the no miracles argument for scientific realism is too quick. Proponents of that argument regard the empirical success of today's scientific theories as evidence of their truth. But the history of science shows that empirically successful theories have often turned out to be false. So how do we know that the same fate will not befall today's theories? How do we know that the atomic theory of matter, for example, will not go the same way as the phlogiston theory? Once we pay due attention to the history of science, argue the anti-realists, we see that the inference from empirical success to theoretical truth is rather shaky. The rational attitude towards the atomic theory is thus one of agnosticism—it may be true or it may not. We just do not know, say the anti-realists.

This is a powerful counter to the no miracles argument, but it is not decisive. Realists have responded by modifying the argument in two ways. The first modification is to claim that a theory's empirical success is evidence that it is approximately true, rather than precisely true. This weaker claim is less vulnerable to counter examples from the history of science. It is also more modest: it allows the realist to admit that today's scientific theories may not be correct down to every last detail, while still holding that they are broadly on the right lines. And as we have seen, the realist needs the notion of approximate truth anyway, to account for idealized models. The second modification of the no miracles argument involves refining the notion of empirical success. Some realists hold that empirical success is not just a matter of fitting the known data, but also allowing us to predict *new* observations that were previously unknown. Relative to this more stringent criterion of empirical success, it is less easy to find historical examples of empirically successful theories which later turned out to be false.

Whether these refinements can save the no miracles argument is debatable. They certainly reduce the number of historical counterexamples, but not to zero. One that remains is the wave theory of light, first put forward by Christian Huygens in 1690. According to this theory, light consists of wave-like vibrations in an invisible medium called the ether, which was supposed to permeate the whole universe. (The rival to the wave theory was the particle theory of light, favoured by Newton, which held that light consists of very small particles emitted by the light source.) The wave theory was not widely accepted until the French physicist Auguste Fresnel formulated a mathematical version of the theory in 1815, and used it to predict some surprising new optical phenomena. Optical experiments confirmed Fresnel's predictions, convincing many 19th-century scientists that the wave theory of light must be true. But modern physics tells us the theory is not true: there is no such thing as the ether, so light doesn't consist of vibrations in it. Again, we have an example of a false but empirically successful theory.

The important feature of this example is that it tells against even the modified version of the no miracles argument. For Fresnel's theory *did* make novel predictions, so qualifies as empirically successful even relative to the stricter notion of empirical success. And it is hard to see how Fresnel's theory can be called 'approximately true', given that it was based around the idea of the ether, which does not exist. Whatever exactly it means for a theory to be approximately true, a necessary condition is surely that the entities the theory talks about really do exist. In short, Fresnel's theory was empirically successful even according to a strict understanding of this notion, but was not even approximately true. The moral of the story, say anti-realists, is that we should not assume that modern scientific theories are even roughly on the right lines, just because they are so empirically successful.

The status of the no miracles argument is thus an open question. On the one hand, the argument is open to serious challenge from the history of science. On the other hand, there is something intuitively compelling about the argument. It really is hard to accept that atoms and electrons might not exist when one considers the amazing success of theories which postulate these entities. But as history shows, we should be cautious about assuming that our current scientific theories are true, however well they fit our data. Many scientists have assumed that in the past and been proved wrong.

The observable/unobservable distinction

Central to the debate between realism and anti-realism is the distinction between what is observable and what is not. So far we have simply taken this distinction for granted—tables and chairs are observable, atoms and electrons are not. But in fact the distinction is quite philosophically problematic. Indeed, one of the main arguments for scientific realism says that it is not possible to draw the distinction in a principled way.

Why should this be an argument for scientific realism? Because anti-realists hold that science cannot give us knowledge of unobservable reality, which presumes that there is a clear distinction between what can be observed and what cannot. If it turns out that this division cannot be drawn satisfactorily, then anti-realism is obviously in trouble. That is why scientific realists are often keen to emphasize the problems associated with the observable/unobservable distinction.

One such problem concerns the relation between observation and detection. Entities such as electrons are not observable in the ordinary sense, but their presence can be detected using special pieces of apparatus called particle detectors. The simplest particle detector is the cloud chamber, which consists of a closed container filled with air that has been saturated with water vapour (see Figure 5). When charged particles such as electrons pass through the chamber, they collide with neutral atoms in the air, converting them into ions; water vapour condenses around these ions causing liquid droplets to form, which can be seen with the naked eye. We can follow the path of an electron through the cloud chamber by watching the tracks of these liquid droplets. Does this mean that electrons can be observed after all? Most philosophers would say no: cloud chambers allow us to detect electrons, not observe them directly. In much the same way, high-speed jets can be detected by the vapour trails they leave behind, but watching these trails is not observing the jet. But is it always clear how to distinguish observing from detecting?

In a well-known defence of scientific realism from the early 1960s, Grover Maxwell posed the following problem for the anti-realist. Consider the following sequence of events: looking at something with the naked eye, looking at something through a window, looking at something through a pair of strong glasses, looking at something through binoculars, looking at something through a low-powered microscope, looking at something through a

5. Cloud chamber.

high-powered microscope, and so on. Maxwell argued that these events lie on a smooth continuum. So how do we decide which count as observing and which not? Can a biologist observe micro-organisms with their high-powered microscope, or can they only detect their presence in the way that a physicist can detect the presence of electrons in a cloud chamber? If something can be only be seen with the help of sophisticated scientific instruments, does it count as observable or unobservable? How sophisticated can the instrumentation be, before we have a case of detecting rather than observing? There is no principled way of answering such questions, Maxwell argued, so the anti-realist's attempt to classify entities as either observable or unobservable fails.

Maxwell's argument is bolstered by the fact that scientists themselves sometimes talk about 'observing' particles with the help of sophisticated bits of apparatus. In the philosophical literature, electrons are usually taken as paradigm examples of unobservable entities, but scientists are often perfectly happy to talk about 'observing' electrons using particle detectors. Of course, this does not prove that the philosophers are wrong and that electrons are observable after all, for the scientists' talk is probably best regarded as a *façon de parler*. Similarly, the fact that scientists talk about having 'experimental proof' of a theory does not mean that experiments can really prove theories to be true, as we saw in Chapter 2. Nonetheless, if there really is a philosophically important observable/unobservable distinction, as anti-realists maintain, it is odd that it corresponds so badly with the way scientists themselves speak.

Maxwell's arguments are powerful but not decisive. Bas van Fraassen, a leading contemporary anti-realist, claims that Maxwell's arguments only show 'observable' to be a vague term. A vague term is one that has borderline cases. 'Bald' is an example. Since hair loss comes in degrees, there are many men of whom it's hard to say whether they are bald or not. But van Fraassen points out that vague terms are perfectly usable, and can mark genuine distinctions in the world. No one would argue that the distinction between bald and hirsute men is unreal simply because 'bald' is vague. Certainly, if we attempt to draw a sharp dividing line it will be arbitrary. But since there are clear-cut cases of men who are bald and men who are not, the lack of a sharp dividing line doesn't matter. Precisely the same applies to 'observable', according to van Fraassen. There are clear-cut cases of entities that can be observed, for example chairs, and entities that cannot, for example quarks. Maxwell's argument highlights the fact that there are also borderline cases, where we are unsure whether the entities in question can be observed or only detected. So if we try to draw a sharp dividing line, it will inevitably be somewhat arbitrary. But as with baldness, this does not show that the observable/unobservable distinction is unreal.

How strong an argument is this? Van Fraassen is probably right that the existence of borderline cases, and the consequent impossibility of drawing a sharp boundary without arbitrariness, does not show the observable/unobservable distinction to be unreal. To that extent, his argument against Maxwell succeeds. However, it is one thing to show that there is a real distinction between what is observable and what is not, and another to show that the distinction merits the significance that anti-realists accord it. Even if we grant van Fraassen his point that there are clear cases of unobservable entities, and that that is enough for the anti-realist to get on with, an argument is still needed for why knowledge of unobservable reality is impossible.

The underdetermination argument

One such argument focuses on the relation between scientists' empirical data and their theories. Anti-realists emphasize that the empirical data to which scientific theories are responsible consist of facts about observable entities and processes. To illustrate, consider again the kinetic theory of gases, which says that any sample of gas consists of molecules in motion. Since these molecules are unobservable, we obviously cannot test the theory by directly observing various samples of gas. Rather, we need to deduce from the theory some statement that can be directly tested, which will invariably be about observable entities. As we saw, the kinetic theory implies that a sample of gas at constant pressure will expand when heated. This statement can be directly tested, by observing the readings on the relevant pieces of apparatus in a laboratory. This example illustrates a general truth: facts about observable phenomena provide the ultimate data for theories that posit unobservable entities and processes.

Anti-realists then argue that the empirical data 'underdetermine' the theories scientists put forward on their basis. What does this mean? It means that the data can in principle be explained by many different, mutually incompatible, theories. In the case of the kinetic theory, anti-realists will say that *one* possible explanation of the empirical data is that gases contain large numbers of molecules in motion, as the kinetic theory says. But they will insist that there are other possible explanations too, which conflict with the kinetic theory. So according to anti-realists, scientific theories which posit unobservable entities are *underdetermined* by the empirical data—there will always be a number of competing theories which can account for the data equally well.

It is easy to see why the underdetermination argument supports an anti-realist view of science. Suppose a scientist believes a given theory about unobservable entities, on the grounds that it can explain a large range of empirical data. However if the data can equally be accounted for by alternative theories, which are mutually incompatible, then the scientist's confidence seems misplaced. For what reason does the scientist have to prefer their theory to one of the alternatives? Underdetermination leads naturally to the anti-realist conclusion that agnosticism is the rational attitude to take towards theories about the unobservable region of reality.

But is it true that the empirical data can always be explained by multiple theories, as anti-realists maintain? Realists typically reply that this is true only in a trivial sense. In principle, there will always be more than one possible explanation of a given set of observations. But, say the realists, it does not follow that all of these possible explanations are equally good. One of the theories might be simpler than the others, for example, or might fit better with theories from another area of science, or might postulate fewer hidden causes. Once we acknowledge that there are criteria for theory choice that go beyond mere compatibility with the data, the problem of underdetermination is defused, according to realists.

This line of thought is bolstered by the fact that there are relatively few real cases of underdetermination in the history of science. If the empirical data can always be explained by many different theories, as anti-realists maintain, surely we should expect to find scientists in near perpetual disagreement with one another? But that is not what we find. Indeed when we inspect the historical record, the situation is almost exactly the reverse of what the underdetermination argument would lead us to expect. Far from scientists being faced with a large number of alternative explanations of their data, they often have difficulty finding even *one* theory that fits the data adequately. This lends support to the realist view that underdetermination is merely a philosopher's worry, with little relation to actual scientific practice.

Anti-realists are unlikely to be impressed by this response. After all, philosophical worries are still genuine ones, even if their practical implications are few. Moreover, the suggestion that criteria such as simplicity can be used to adjudicate between competing theories immediately invites the awkward question of why simpler theories should be thought more likely to be true; we touched on this issue in Chapter 2.

Anti-realists typically grant that underdetermination can be eliminated in practice by using criteria such as simplicity to discriminate between competing explanations of our data. But they deny that such criteria are reliable indicators of the truth.

Simpler theories may be more convenient to work with, but they are not intrinsically more probable than complex ones. So the underdetermination argument stands: in principle there are always multiple explanations of the empirical data, we have no way of knowing which is true, so knowledge of unobservable reality cannot be had.

However the story does not end here; there is a further realist comeback. Realists accuse anti-realists of applying the underdetermination argument selectively. If the argument is applied consistently, it rules out not only knowledge of the unobservable world, but also knowledge of much of the *observable* world, say the realists. To understand why realists say this, notice that many things that are observable never actually get observed. For example, the vast majority of living organisms on the planet never get observed by humans, but they are clearly observable. Or think of an event such as a large meteorite hitting the earth. No one has ever witnessed such an event, but it is clearly observable. It just so happens that no human was ever in the right place at the right time. Only a small fraction of what is observable actually gets observed.

The key point is this. Anti-realists claim that the unobservable region of reality lies beyond the limits of scientific knowledge. So they allow that we can have knowledge of objects and events that are observable but unobserv*ed*. But theories about the unobserved are no less underdetermined by our data than theories about the unobservable. For example, suppose a scientist advances the hypothesis that a meteorite struck the moon in 1987. They cite various pieces of observational data to support this hypothesis, e.g. that satellite pictures of the moon show a large crater that wasn't there before 1987. However, this data can in principle be explained by alternative hypotheses—perhaps a volcanic eruption caused the crater, or an earthquake. Or perhaps the camera that took the satellite pictures was faulty, and there is no crater at all. So the scientist's hypothesis is underdetermined by the data, even though the hypothesis is about a perfectly observable event—a meteorite striking the moon. If we apply the underdetermination argument consistently, say realists, we are forced to conclude that science can only give us knowledge of things that have actually been observed.

This conclusion is not one that many philosophers of science would accept. For much of what science tell us concerns things that have not been observed—think of ice ages, dinosaurs, and continental drift. To say that knowledge of the unobserved is impossible is to say that most of what passes for scientific knowledge is not really knowledge at all. Scientific realists take this to show that the underdetermination argument must be wrong. Since science clearly does give us knowledge of the unobserved, despite the fact that theories about the unobserved are underdetermined by our data, it follows that underdetermination is no barrier to knowledge. So the fact that our theories about the unobservable are also underdetermined does not mean that science cannot give us knowledge of the unobservable.

In effect, realists who argue this way are saying that the underdetermination problem is simply Hume's problem of induction in disguise. Underdetermination means that the data can be accounted for by alternative theories. But this is effectively just to say that the data do not entail the theory: the inference from data to theory is non-deductive. Whether the theory is about unobservable entities, or about observable but unobserved entities, makes no difference—the logic of the situation is the same in both cases. Of course, showing that the underdetermination argument is just a version of the problem of induction does not mean that it can be ignored. But it does mean that there is no *special* difficulty about unobservable entities. Therefore the anti-realist position is ultimately arbitrary, say the realists. Whatever problems there are in understanding how science can give us knowledge of atoms and electrons are equally problems for understanding how science can give us knowledge of ordinary macroscopic objects.

Chapter 5
Scientific change and scientific revolutions

Scientific ideas change fast. Pick virtually any scientific discipline you like, and you can be sure that the prevalent theories in the discipline will be different from those of fifty years ago, and very different from those of 100 years ago. Compared with other areas of intellectual endeavour, science is a rapidly changing activity. A number of interesting philosophical questions centre on the issue of scientific change. Is there a discernible pattern to the way scientific ideas change over time? When scientists abandon their existing theory in favour of a new one, how should we explain this? Are later scientific theories objectively better than earlier ones?

Most modern discussion of these questions takes off from the work of Thomas Kuhn, an American historian and philosopher of science. In 1963 Kuhn published a book called *The Structure of Scientific Revolutions*, which had an enormous influence on subsequent philosophy of science. The impact of Kuhn's ideas has also been felt in academic disciplines such as sociology and anthropology, and in the intellectual culture at large. (*The Guardian* newspaper included *The Structure of Scientific Revolutions* in its list of the 100 most influential books of the 20th century.) To understand why Kuhn's ideas caused such a stir, we need to look briefly at the state of philosophy of science prior to the publication of his book.

Logical empiricist philosophy of science

The dominant philosophical movement in the English-speaking world in the post-war period was *logical empiricism*. The original logical empiricists were a loosely knit group of philosophers, logicians, and scientists who gathered in Vienna and Berlin in the 1920s and early 1930s. (Carl Hempel, whom we met in Chapter 3, was closely associated with the group, as was Karl Popper.) Fleeing persecution by the Nazis, most of the logical empiricists emigrated to the United States, where they and their followers exerted a powerful influence on academic philosophy until about the mid-1960s, by which time the movement had begun to disintegrate.

The logical empiricists had a high regard for the natural sciences and also for mathematics and logic. The early years of the 20th century witnessed exciting scientific advances, particularly in physics, which impressed them tremendously. One of their aims was to make philosophy itself more 'scientific', in the hope that this would allow similar advances to be made in philosophy. What impressed the logical empiricists about science was its apparent objectivity. Unlike in other fields, where much turned on the subjective opinion of enquirers, scientific questions could be settled in a fully objective way, they believed. Techniques such as experimental testing allowed a scientist to compare their theory directly with the facts, and thus reach an informed, unbiased decision about the theory's merits. Science for the logical empiricists was thus a paradigmatically *rational* activity, the surest route to the truth that there is.

Despite the high esteem in which they held science, the logical empiricists paid little attention to the history of scientific ideas. This was primarily because they drew a sharp distinction between what they called the 'context of discovery' and the 'context of justification'. The context of discovery refers to the actual historical process by which a scientist arrives at a given theory. The context of justification refers to the means by which the scientist tries to justify the theory once they already have it—which includes testing the theory, searching for relevant evidence, and comparing it with rival theories. The logical empiricists believed that the former was a subjective, psychological process which wasn't governed by precise rules, while the latter was an objective matter of logic. Philosophers of science should confine themselves to studying the latter, they argued.

An example can help illustrate this idea. In 1865 the German chemist Kekulé discovered that the benzene molecule has a hexagonal structure. Apparently, he hit on the hypothesis of a hexagonal structure after a dream in which he saw a snake trying to bite its own tail (see Figure 6). Of course, Kekulé then had to

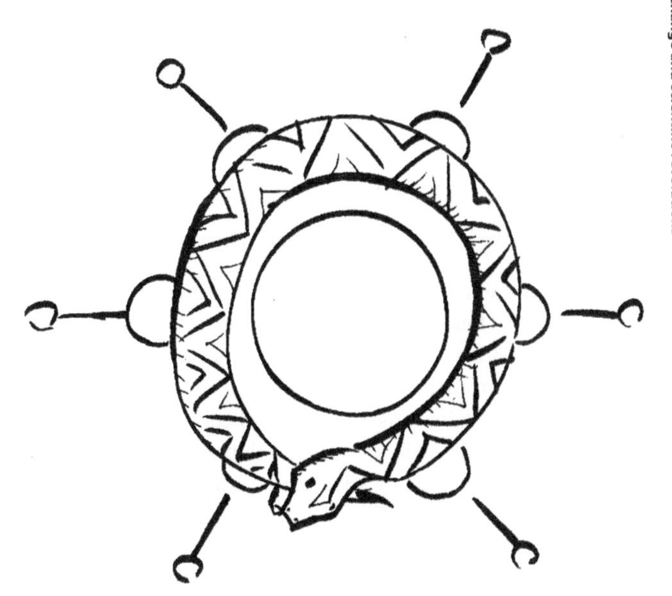

6. Kekulé arrived at the hypothesis of the hexagonal structure of benzene after a dream in which he saw a snake trying to bite its own tail.

test his hypothesis scientifically before it could be accepted. This is an extreme example, but it shows that scientific hypotheses can be arrived at in the most unlikely of ways—they are not always the product of careful, systematic thought. The logical empiricists held that it makes no difference how a hypothesis is arrived at initially. What matters is how it is tested once it is already there—for it is this that makes science a rational activity.

Another theme in logical empiricist philosophy of science was the distinction between theories and observational facts; this is related to the observable/unobservable distinction discussed in Chapter 4. The logical empiricists believed that disputes between rival scientific theories could be solved in a fully objective way—by comparing the theories directly with the 'neutral' observational facts, which all parties could accept. How exactly this set of neutral facts should be characterized was a matter of debate among the logical empiricists, but they were adamant that it existed. Without a clear distinction between theories and observational facts the rationality and objectivity of science would be compromised, and they were resolute in their belief that science was rational and objective.

Kuhn's theory of scientific revolutions

Kuhn was a historian of science by training, and firmly believed that philosophers had much to learn from studying the history of science. Insufficient attention to the history of science had led the logical empiricists to form an inaccurate and naive picture of the scientific enterprise, he maintained. As the title of his book indicates, Kuhn was especially interested in scientific revolutions—periods of great upheaval when existing scientific ideas are replaced with radically new ones. Examples include the Copernican revolution in astronomy, the Einsteinian revolution in physics, and the Darwinian revolution in biology. Each of these revolutions led to a fundamental change in the scientific worldview—the overthrow of an existing set of ideas by a completely different set.

Of course, scientific revolutions happen relatively infrequently—most of the time any given science is not in a state of revolution. Kuhn coined the term 'normal science' to describe the ordinary day-to-day activities that scientists engage in when their discipline is not undergoing revolutionary change. Central to Kuhn's account of normal science is the concept of a *paradigm*. A paradigm consists of two main components: first, a set of fundamental theoretical assumptions which all members of a scientific community accept; secondly, a set of 'exemplars' or particular scientific problems which have been solved by means of those theoretical assumptions, and which appear in the textbooks of the discipline in question. But a paradigm is more than just a theory (though Kuhn sometimes uses the words interchangeably). When scientists share a paradigm they do not just agree on certain scientific propositions, they agree also on how future research in their field should proceed, on which problems are the pertinent ones to tackle, on what the appropriate methods for solving those problems are, and on what an acceptable solution of the problems would look like. In short, a paradigm is an entire scientific outlook—a constellation of shared assumptions, beliefs, and values which unite a scientific community and allow normal science to take place.

What exactly does normal science involve? According to Kuhn it is primarily a matter of *puzzle-solving*. However successful a paradigm is, it will always encounter certain problems—phenomena which it cannot easily accommodate, or mismatches between the theory's predictions and the experimental facts. The job of the normal scientist is to try to eliminate these minor puzzles while making as few changes as possible to the paradigm. So normal science is a conservative activity—its practitioners are not trying to make any earth-shattering discoveries, but rather just to develop and extend the existing paradigm. In Kuhn's words, 'normal science does not aim at novelties of fact or theory, and when successful finds none'. Above all, Kuhn stressed that normal scientists are not trying to *test* the paradigm. On the contrary, they accept the paradigm unquestioningly, and conduct their research within the limits it sets. If a normal scientist gets an experimental result which conflicts with the paradigm, they will usually assume that their experimental technique is faulty, not that the paradigm is wrong.

Typically a period of normal science lasts many decades, sometimes even centuries. During this time scientists gradually articulate the paradigm—fine-tuning it, filling in details, and extending its range of application. But over time *anomalies* are discovered—phenomena which simply cannot be reconciled with the paradigm, however hard scientists try. When anomalies are few they tend to just get ignored. But as anomalies accumulate, a burgeoning sense of crisis envelops the scientific community. Confidence in the existing paradigm breaks down, and the process of normal science grinds to a halt. This marks the beginning of a period of 'revolutionary science' as Kuhn calls it. During such periods, fundamental scientific ideas *are* up for grabs. A variety of alternatives to the old paradigm are proposed, and eventually a new paradigm becomes established. A generation is usually required before all members of the scientific community are won over to the new paradigm—an event which marks the completion of a scientific revolution. The essence of a scientific revolution is thus the shift from an old paradigm to a new one.

Kuhn's characterization of the history of science as long periods of normal science punctuated by occasional scientific revolutions struck a chord with many scholars. A number of examples from the history of science fit Kuhn's description quite well. In the transition from Ptolemaic to Copernican astronomy, for example, or from Newtonian to Einsteinian physics, many of the features that Kuhn describes are present. Ptolemaic astronomers did indeed share a paradigm, based around the theory that the earth is stationary at the centre of the universe, which formed the unquestioned back-drop to their investigations. The same is true of Newtonian physicists in the 18th and 19th centuries, whose paradigm was based around Newton's theory of mechanics and gravitation. And in both cases, Kuhn's account of how an old paradigm gets replaced by a new one applies fairly accurately. There are also scientific revolutions which do not fit the Kuhnian model so neatly—for example the molecular revolution in biology in the 1950s and 1960s. But nonetheless, most people agree that Kuhn's description of the history of science contains much of value.

Why did Kuhn's ideas cause such a storm? Because in addition to his descriptive claims about the history of science, Kuhn advanced some controversial philosophical theses. Ordinarily we assume that when scientists trade their existing theory for a new one, they do so on the basis of evidence. But Kuhn argued that adopting a new paradigm involves a certain act of faith on the part of the scientist. He allowed that a scientist could have good reasons for abandoning an old paradigm for a new one, but he insisted that reasons alone could never rationally *compel* a paradigm shift. 'The transfer of allegiance from paradigm to paradigm', Kuhn wrote, 'is a conversion experience which cannot be forced.' And in explaining why a new paradigm rapidly gains acceptance in the scientific community, Kuhn emphasized the peer pressure of scientists on one another. If a given paradigm has very forceful advocates, it is more likely to win widespread acceptance.

Many of Kuhn's critics were appalled by these claims. For if paradigm shifts work the way Kuhn says, it is hard to see how science can be regarded as a rational activity at all. Surely scientists are meant to base their beliefs on evidence and reason, not on faith and peer pressure? Faced with two competing paradigms, surely the scientist should make an objective comparison of them to determine which has more evidence in its favour? Undergoing a 'conversion experience', or allowing oneself to be persuaded by the most forceful of one's fellow scientists, hardly seems like a rational way to behave. One critic wrote that on Kuhn's account, theory choice in science was 'a matter for mob psychology'.

Kuhn also made some controversial claims about the overall direction of scientific change. According to a widely held view, science progresses towards the truth in a linear fashion, as older incorrect ideas get replaced by newer, correct ones. Later theories are thus objectively better than earlier ones, so scientific knowledge accumulates over time. This linear, cumulative conception of science is popular among laypeople and scientists alike, but Kuhn argued that it is both historically inaccurate and philosophically naive.

For example, he noted that Einstein's theory of relativity is in some respects more similar to Aristotelian than Newtonian physics—so the history of mechanics is not simply a linear progression from wrong to right. Moreover, Kuhn questioned whether the concept of objective truth actually makes sense at all. The idea that there is a fixed set of facts about the world, independent of any particular paradigm, was of dubious coherence, he believed. Kuhn suggested a radical alternative: the facts about the world are paradigm-relative, and thus change when paradigms change. If this suggestion is right, then it makes no sense to ask whether a given theory corresponds to the facts 'as they really are', nor therefore to ask whether it is objectively true. This led Kuhn to espouse a radical form of anti-realism about science.

Incommensurability and the theory-ladenness of data

Kuhn had two main philosophical arguments for these claims. First, he argued that competing paradigms are typically 'incommensurable' with one another. To understand this idea, recall that for Kuhn a scientist's paradigm determines their entire worldview. So when an existing paradigm is replaced by a new one in a scientific revolution, scientists have to abandon the whole conceptual framework which they use to make sense of the world. Indeed Kuhn even claims, somewhat metaphorically, that before and after a paradigm shift scientists 'live in different worlds'.

Incommensurability is the idea that two paradigms may be so different as to render impossible any straightforward comparison of them with each other—there is no common language into which both can be translated. As a result, the proponents of different paradigms 'fail to make complete contact with each other's viewpoints', Kuhn claimed.

This is an interesting if somewhat vague idea. The doctrine of incommensurability stems from Kuhn's belief that scientific concepts derive their meaning from the theory in which they play a role. So to understand Newton's concept of mass, for example, we need to understand the whole of Newtonian theory—concepts cannot be explained independently of the theories in which they are embedded. This idea, which is sometimes called 'holism', was taken very seriously by Kuhn. He argued that the term 'mass' actually meant something different for Newton and Einstein, since the theories in which each embedded the term were so different. This implies that Newton and Einstein were in effect speaking different languages, which obviously complicates the attempt to compare their theories. If a Newtonian and an Einsteinian physicist tried to have a rational discussion, they would end up talking past each other.

Kuhn used the incommensurability thesis both to rebut the view that paradigm shifts are fully 'objective', and to bolster his non-cumulative picture of the history of science. Traditional philosophy of science saw no huge difficulty in choosing between competing theories—one simply makes an objective comparison of them in the light of the available evidence. But this clearly presumes that there is a common language in which both theories can be expressed. If Kuhn is right that proponents of old and new paradigms are quite literally talking past each other, no such simplistic account of paradigm choice can be correct. Incommensurability is equally problematic for the traditional linear picture of scientific history. If old and new paradigms are incommensurable, then it cannot be correct to think of scientific revolutions as the replacement of 'wrong' ideas by 'right' ones. For to call one idea right and another wrong implies the existence of a common framework for evaluating them, which is precisely what Kuhn denies. Incommensurability implies that scientific change, far from being a straightforward progression towards the truth, is in a sense directionless: later paradigms are not better than earlier ones, just different.

Not many philosophers were convinced by Kuhn's incommensurability thesis. Part of the problem was that Kuhn also claimed old and new paradigms to be *incompatible*. This claim is plausible, for if old and new paradigms were not incompatible there would be no need to choose between them. And in many cases the incompatibility is anyway obvious—the Ptolemaic claim that the planets revolve around the earth is obviously incompatible with the Copernican claim that they revolve around the sun. But as Kuhn's critics were quick to point out, if two things are incommensurable then they cannot be incompatible. To see why not, consider the proposition that an object's mass depends on its velocity. Einstein's theory says this proposition is true while Newton's says it is false. But if the doctrine of incommensurability is right, then there is no actual disagreement between Newton and Einstein here, for the proposition means something different for each. Only if the proposition has the *same* meaning in both theories is there a genuine conflict between the two. Since everybody (including Kuhn) agrees that Einstein's and Newton's theories *do* conflict, this is strong reason to regard the incommensurability thesis with suspicion.

In response to this objection, Kuhn moderated his incommensurability thesis somewhat. He argued that partial translation between paradigms could be achieved, so the proponents of old and new paradigms could communicate to some extent: they would not always be talking past each other entirely. But Kuhn continued to maintain that fully objective choice between paradigms was impossible. For in addition to the incommensurability deriving from the lack of a common language, there is also what he called 'incommensurability of standards'. This is the idea that proponents of different paradigms may disagree about what features a good paradigm should have, what problems it should be able to solve, and what an acceptable solution to those problems would look like. So even if they can communicate effectively, they will not be able to reach agreement about whose paradigm is superior. In Kuhn's words, 'each paradigm will be shown to satisfy the criteria that it dictates for itself and to fall short of a few of those dictated by its opponent'.

Kuhn's second philosophical argument was based on an idea known as the 'theory-ladenness' of data. To grasp this idea, suppose you are a scientist trying to choose between two conflicting theories. The obvious thing to do is to look for a piece of data which will decide between them, or to perform a 'crucial experiment' that will settle the matter. But this will only be possible if there exist data which are suitably independent of the theories, in the sense that a scientist could accept the data whichever of the two theories they believed. As we have seen, the logical empiricists believed in the existence of such theory-neutral data, which could provide an objective court of appeal between competing theories. But Kuhn argued that the ideal of theory-neutrality is an illusion—data are invariably contaminated by theoretical assumptions. It is impossible to isolate a set of 'pure' data which all scientists would accept irrespective of their theoretical persuasion, he argued.

The theory-ladenness of data had two important consequences for Kuhn. First, it meant that a dispute between competing paradigms could not be resolved by simply appealing to 'the data' or 'the facts', for what a scientist counts as data, or facts, will depend on which paradigm they accept. Perfectly objective choice between two paradigms is therefore impossible: there is no neutral vantage-point from which to assess the claims of each.

Secondly, the very idea of objective truth is called into question. To be objectively true, a theory must correspond to the facts, but the idea of such a correspondence makes little sense if the facts themselves are infected by our theories. This is why Kuhn was led to the radical view that truth itself is relative to a paradigm.

Why did Kuhn think that all data are theory-laden? His writings are not totally clear on this point, but at least two lines of argument are discernible. The first is the idea that perception is heavily conditioned by background beliefs—what we see depends in part on what we believe. So a trained scientist looking at a sophisticated piece of apparatus in a laboratory will see something different from what a layperson sees, for the scientist obviously has many beliefs about the apparatus that the layperson lacks. There are a number of psychological experiments which apparently show that perception is sensitive in this way to background belief—though the correct interpretation of these experiments is a contentious matter. Secondly, scientists' experimental and observational reports are often couched in highly theoretical language. For example, a scientist might report the outcome of an experiment by saying 'an electric current is flowing through the copper rod'. But this data report is obviously laden with a large amount of theory. It would not be accepted by a scientist who did not hold standard beliefs about electric currents, so it is clearly not theory-neutral.

Philosophers are divided over the merits of these arguments. On the one hand, many agree with Kuhn that pure theory-neutrality is unattainable. The logical empiricist ideal of a class of data statements totally free of theoretical commitment is rejected by most contemporary philosophers—not least because no one has succeeded in saying what such statements would look like. But this need not compromise the objectivity of paradigm shifts altogether. Suppose that a Ptolemaic and a Copernican astronomer are engaged in a debate about whose theory is superior. In order for them to debate meaningfully, there needs to be some astronomical data they can agree on. But why should this be a problem? Surely they can agree about the relative position of the earth and the moon on successive nights, for example, or the time at which the sun rises? Obviously, if the Copernican insists on describing the data in a way that presumes the truth of the heliocentric theory, the Ptolemaist will object. But there is no reason why the Copernican should do that. Statements like 'on May 14th the sun rose at 7.10 am' can be agreed on by a scientist whether they believe the geocentric or the heliocentric theory. Such statements are sufficiently theory-neutral to be acceptable to proponents of both paradigms, which is what matters.

What about Kuhn's rejection of objective truth? Few philosophers have followed Kuhn's lead here. Part of the problem is that, like many who reject the concept of objective truth, Kuhn failed to articulate a viable alternative. The radical view that truth is paradigm-relative is ultimately hard to make sense of. For like all relativist doctrines, it faces a critical problem. Consider the question: is the claim that truth is paradigm-relative *itself* objectively true or not? If the proponent of relativism answers 'yes', then they have admitted that the concept of objective truth does make sense and thus contradicted themselves. If they answer 'no', then they have no grounds on which to argue with someone who disagrees and says that, in their opinion, truth is *not* paradigm-relative. Not all philosophers regard this argument as completely fatal to relativism, but it does suggest that abandoning the concept of objective truth is easier said than done. Kuhn certainly raised some telling objections to the traditional view that the history of science is simply a linear progression to the truth, but the relativist alternative he offered in its place is not easy to accept.

Kuhn and the rationality of science

The Structure of Scientific Revolutions is written in a radical tone. Kuhn gives the impression of wanting to replace standard philosophical ideas about theory change in science with a radically different conception. His doctrines of paradigm shifts, incommensurability, and the theory-ladenness of data seem wholly at odds with the logical empiricist view of science as rational, objective, and cumulative. With some justification, Kuhn's readers took him to be saying that science is a largely non-rational activity, characterized by dogmatic adherence to a paradigm in normal periods, and sudden 'conversion experiences' in revolutionary periods.

But Kuhn himself was unhappy with this interpretation of his work. In a Postscript to the second edition of *Structure* published in 1970, and in subsequent writings, Kuhn moderated his tone considerably, distancing himself from the more radical views that he had seemed to endorse. He was not trying to cast doubt on the rationality of science, he argued, but rather to offer a more realistic, historically accurate picture of how science actually develops. By neglecting the history of science, the logical empiricists had been led to a simplistic account of how science works, and Kuhn's aim was to provide a corrective. He was not trying to show that science was irrational, but rather to provide a better account of what scientific rationality involves.

Some commentators regard Kuhn's Postscript as an about-turn—a retreat from his original position rather than a clarification of it. Whether this is a fair assessment is not a question we will consider here. But the Postscript did bring to light one important issue. In rebutting the charge that he had portrayed science as non-rational, Kuhn made the famous claim that there is 'no algorithm' for theory choice in science. What does this mean? An algorithm is a set of rules which allows us to compute the answer to a particular question. For example, an algorithm for multiplication is a set of rules which when applied to any two numbers, tells us their product. So an algorithm for theory choice is a set of rules which, when applied to two competing theories, would tell us which to choose. Much traditional philosophy of science was committed, implicitly or explicitly, to the existence of such an algorithm. The logical empiricists often wrote as if, given a set of data and two competing theories, the 'principles of scientific method' could be used to determine which theory was superior. This idea was implicit in their belief that although discovery was a matter of psychology, justification was a matter of logic.

Kuhn's insistence that there is no algorithm for theory-choice in science is probably correct. Certainly no one has ever succeeded in producing such an algorithm. Lots of philosophers and scientists have made plausible suggestions about what to look for in theories—simplicity, broadness of scope, close fit with the data, and so on. But these suggestions fall short of providing a true algorithm, as Kuhn knew well. For one thing, there may be trade-offs: theory A may be simpler than theory B, but B may fit the data more closely. So an element of subjective judgement, or scientific common sense, will often be needed to decide between competing theories. Seen in this light, Kuhn's suggestion that the adoption of a new paradigm involves a certain act of faith does not seem quite so radical, and likewise his emphasis on the persuasiveness of a paradigm's advocates in determining its chance of winning over the scientific community.

The 'no algorithm' idea lends support to the view that Kuhn's account of paradigm shifts is not an assault on the rationality of science. For we can read Kuhn instead as rejecting a certain conception of rationality. The logical empiricists believed, in effect, that there *must* be an algorithm for theory-choice on pain of scientific change being irrational. This is not a crazy view: many paradigm cases of rational action do involve rules or algorithms. For example, if you want to decide whether a good is cheaper in England or Japan, you apply an algorithm for converting pounds into yen; any other way of trying to decide the matter is irrational. Similarly, if a scientist is trying to decide between two competing theories, it is tempting to think that the only rational way to proceed is to apply an algorithm for theory-choice. So if it turns out that there is no such algorithm, as seems likely, we have two options. Either we can conclude that scientific change is irrational *or* that the conception of rationality at work is too demanding. In his later writings Kuhn endorses the latter option. The moral of his story is not that paradigm shifts are irrational, but rather that a more relaxed, pragmatic concept of rationality is required to make sense of them.

Kuhn's legacy

Despite their controversial nature, Kuhn's ideas transformed philosophy of science. In part this is because Kuhn called into question many assumptions that had traditionally been taken for granted, forcing philosophers to confront them, and in part because he drew attention to a range of issues that traditional philosophy of science had simply ignored. After Kuhn, the idea that philosophers could afford to ignore the history of science appeared increasingly untenable, as did the idea of a sharp dichotomy between the contexts of discovery and justification. Contemporary philosophers of science pay much greater attention to the historical development of science than did their pre-Kuhnian ancestors. Even those unsympathetic to Kuhn's more radical ideas would accept that in these respects his influence has been positive.

Another important impact of Kuhn's work was to focus attention on the social context in which science takes place, something that traditional philosophy of science ignored. Science for Kuhn is an intrinsically social activity: the existence of a scientific community, bound together by allegiance to a shared paradigm, is a prerequisite for the practice of normal science. Kuhn also paid considerable attention to how science is taught in schools and universities, how young scientists are initiated into the scientific community, how scientific results are published, and other such 'sociological' matters. Not surprisingly, Kuhn's ideas have been influential among sociologists of science. In particular, a movement known as the 'strong programme' in the sociology of science, which emerged in Britain in the 1970s and 1980s, owes much to Kuhn.

The strong programme was based around the idea that science should be viewed as a product of the society in which it is practised. Strong programme sociologists took this idea literally: they held that scientists' beliefs were in large part socially determined. So to explain why a scientist believes a given theory, for example, they would cite aspects of the scientist's social and cultural background. The scientist's own reasons for believing the theory were never explanation enough, they maintained. The strong programme borrowed a number of themes from Kuhn, including the theory-ladenness of data, the view of science as an essentially social activity, and the idea that there is no objective algorithm for theory-choice. But strong programme sociologists were more radical than Kuhn, and less cautious. They openly rejected the notions of truth and rationality, which they regarded as ideologically suspect, and viewed traditional philosophy of science with great suspicion. This led to a certain amount of tension between philosophers and sociologists of science, which continues to this day.

Further afield, Kuhn's work has played a role in the rise of *social constructionism* in the humanities and social sciences. Social constructionism is the idea that certain phenomena, e.g. racial categories, are 'social constructs', as opposed to having an objective mind-independent existence. Given Kuhn's emphasis on the social context of science, and his rejection of the idea that scientific theories 'correspond to the objective facts', it is easy to see why he could be read as saying that science is a 'social construct'. However there is a certain irony here. For proponents of the idea that science is a 'social construct' have typically had an anti-scientific attitude, often objecting to the authority accorded to science in modern society. But Kuhn himself was strongly pro-science. Like the logical empiricists, he regarded modern science as a hugely impressive intellectual achievement. His doctrines of paradigm shifts, normal and revolutionary science, incommensurability, and theory-ladenness were not intended to undermine or criticize the scientific enterprise, but rather to help us understand it better.

Chapter 6
Philosophical problems in physics, biology, and psychology

The issues we have studied so far—inference, explanation, realism, and scientific change—belong to what is called 'general philosophy of science'. These issues concern the nature of scientific investigation in general, rather than pertaining specifically to chemistry, say, or geology. However, there are also interesting philosophical questions that are specific to particular sciences—they belong to what is called 'philosophy of the special sciences'. These questions usually depend partly on philosophical considerations and partly on empirical facts, which is what makes them so interesting. In this chapter we examine three such questions, one each from physics, biology, and psychology.

Leibniz versus Newton on absolute space

Our first topic is a debate between Gottfried Leibniz (1646–1716) and Isaac Newton (1642–1727), two of the outstanding scientific intellects of the 17th century, concerning the nature of space and time. In his *Principles of Natural Philosophy*, Newton defended what is called an 'absolutist' conception of space. According to this view, space has an 'absolute' existence over and above the spatial relations between objects. Newton thought of space as a three-dimensional container into which God placed the material universe at creation. This implies that space existed before there were any material objects, just as a container like a cereal box exists before any pieces of cereal are put inside. The only difference between space and ordinary containers like cereal boxes, according to Newton, is that the latter have finite dimensions whereas space extends infinitely in every direction.

Leibniz strongly disagreed with the absolutist view of space, and with much else in Newton's philosophy. He argued that space consists simply of the totality of spatial relations between material objects. Example of spatial relations are 'above', 'below', 'to the left of', and 'to the right of'—they are relations that material objects bear to each other. This 'relationist' conception of space implies that before there were any material objects, space did not exist. Leibniz argued that space came into existence *when* God created the material universe; it did not exist beforehand, waiting to be filled up with material objects. So space is not usefully thought of as a container, nor indeed as an entity of any sort. Leibniz's view can be understood in terms of an analogy. A legal contract consists of a relationship between two parties—the buyer and seller of a house, for example. If one of the parties dies, then the contract ceases to exist. So it would be crazy to say that the contract has an existence independently of the relationship between buyer and seller—the contract just *is* that relationship. Similarly, space is nothing over and above the spatial relations between objects.

Newton's main reason for introducing the concept of absolute space was to distinguish relative from absolute motion. Relative motion is the movement of one object with respect to another. So far as relative motion is concerned, it makes no sense to ask whether an object is 'really' moving or not—we can only ask whether it is moving with respect to something else. To illustrate, imagine two joggers running in tandem along a straight road. Relative to a bystander on the road side, both are in motion—they are getting further away by the moment. But relative to each other, the joggers are not in motion—their relative positions remain exactly the same, so long as they keep jogging in the same direction at the same speed. So an object may be in relative motion with respect to one thing but be stationary with respect to another.

Newton believed that as well as relative motion, there is also absolute motion. Common sense supports this view. Imagine two objects in relative motion—say a hang-glider and an observer on the earth. Now relative motion is symmetric: just as the hang-glider is in motion relative to the observer on the earth, so the observer is in motion relative to the hang-glider. But intuitively, it surely makes sense to ask whether the observer or the hang-glider is 'really' the one moving? If so, then we need the concept of absolute motion.

But what exactly is absolute motion? According to Newton, it is the motion of an object *with respect to absolute space*. Newton thought that at any time, every object has a particular location in absolute space. If an object changes its location in absolute space over time then it is in absolute motion; otherwise, it is at absolute rest. So we need to think of space as an entity, over and above the relations between material objects, in order to distinguish relative from absolute motion. Notice that Newton's reasoning rests on an important assumption, namely that all motion has got to be relative to *something*. Relative motion is motion relative to other material objects; absolute motion is motion relative to absolute space itself. So in a sense, even absolute motion is 'relative' for Newton. In effect, Newton is assuming that being in motion, whether absolute or relative, cannot be a 'brute fact' about an object; it can only be a fact about the object's relations to something else. That something else can either be another material object, or it can be absolute space.

Leibniz accepted that there was a difference between relative and absolute motion, but denied that the latter should be explained as motion with respect to absolute space. For he regarded the concept of absolute space as incoherent. He had a number of arguments for this view, many of which were theological. From a philosophical viewpoint, Leibniz's most interesting argument was that absolute space conflicts with what he called the *principle of the identity of indiscernibles* (PII). Since Leibniz regarded this principle as indubitably true, he rejected the concept of absolute space.

PII says that if two objects are indiscernible then they are identical, i.e. they are really one and the same object. To call two objects indiscernible is to say that no difference at all can be found between them—they have exactly the same attributes. So if PII is true, then any two genuinely distinct objects must differ in some attribute—otherwise they would be one, not two. PII is intuitively quite compelling. It is certainly not easy to find an example of two distinct objects which share *all* their attributes—even two mass-produced factory goods will normally differ in innumerable ways. Whether PII is true in general is a complex question that philosophers still debate; the answer depends in part on exactly what counts as an 'attribute', and in part on difficult issues in quantum physics. But for the moment our concern is the use to which Leibniz puts the principle.

Leibniz uses two thought experiments to reveal a conflict between Newton's theory of absolute space and PII. Firstly, Leibniz asks us to imagine two different universes, both containing exactly the same objects. In universe one, each object occupies a particular location in absolute space. In universe two, each object has been shifted to a different location in absolute space, two kilometres to the east (for example). There would be no way of telling these two universes apart. For we cannot observe the position of an object in absolute space, as Newton himself admitted. All we can observe are the positions of objects *relative to each other*, and these are the same in both universes. No observations or experiments could ever reveal whether we lived in universe one or two.

Leibniz's second thought experiment is similar. Recall that for Newton, some objects are moving through absolute space while others are at rest. This means that at each moment, every object has a definite absolute velocity. (Velocity is speed in a given direction, so an object's absolute velocity is the speed at which it moves through absolute space in a specified direction.) Now imagine two different universes, both containing exactly the same objects. In universe one, each object has a particular absolute velocity. In universe two, the absolute velocity of each object has been boosted by a fixed amount, say 300 kilometres per hour in a specified direction. Again, we could never tell these two universes apart. For we cannot observe how fast an object is moving with respect to absolute space, as Newton admitted, but only observe how fast objects are moving *relative to each other*—and these are the same in both universes.

In each of these thought experiments, Leibniz describes two universes which by Newton's own admission we could never tell apart—they are perfectly indiscernible. But by PII, this means that the two universes are actually one. So if PII is true, then Newton's theory has a false consequence: it implies that there are two things when there is only one. Therefore Newton's theory is false, Leibniz argues.

In effect, Leibniz is arguing that absolute space is an empty notion because it makes no observational difference. If neither the location of objects in absolute space nor their velocity with respect to absolute space can ever be detected, why believe in absolute space at all? Leibniz is appealing to the quite reasonable principle that we should only postulate unobservable entities in science if their existence would make a difference that we can detect observationally.

But Newton thought he could show that absolute space *did* have observational effects. This is the point of his famous 'rotating bucket' argument. He asks us to imagine a bucket full of water, suspended on a rope threaded through a hole in its base (see Figure 7). Initially the water is at rest relative to the bucket. The

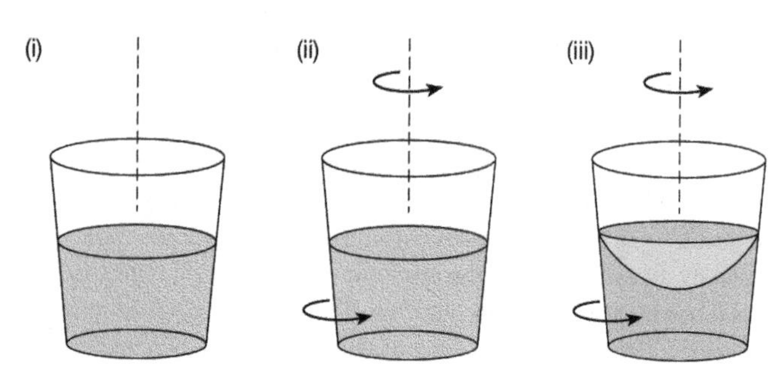

7. Newton's 'rotating bucket' experiment. Stage (i) bucket and water are at rest; stage (ii) bucket rotates relative to water; stage (iii) bucket and water rotate in tandem.

rope is then twisted around a number of times and released. As it uncoils, the bucket starts rotating. At first the water in the bucket stays still, its surface flat; the bucket is then rotating relative to the water. But after a few moments the bucket imparts its motion to the water, and the water begins to rotate in tandem with the bucket; the bucket and water are then at rest relative to each other again. Experience shows that the surface of the water then curves upwards at the sides.

What is causing the surface of the water to rise, Newton asks? Clearly it is something to do with the water's rotation. But rotation is a type of motion, and for Newton an object's motion is always relative to something else. So we must ask: relative to what is the water rotating? Not relative to the bucket, obviously, for the bucket and water are rotating in tandem and are hence at relative rest. Newton argues that the water is rotating *relative to absolute space*, and that this is causing its surface to curve upwards. So absolute space does in fact have observational effects.

You may think there is an obvious gap in Newton's argument. Granted the water is not rotating relative to the bucket, but why conclude that it must be rotating relative to absolute space? The water is rotating relative to the person doing the experiment, and relative to the earth's surface, and relative to the fixed stars, so surely any of these might be causing its surface to rise? However Newton had a simple response to this move. Imagine a universe containing nothing except the rotating bucket. In such a universe, we cannot explain the water's curved surface by appealing to the water's rotation relative to other objects, for there are none, and as before the water is at rest relative to the bucket. Absolute space is the only thing left for the water to be rotating relative to. So we must believe in absolute space on pain of being unable to explain why the water's surface curves.

In effect, Newton is saying that although an object's position and velocity with respect to absolute space can never be detected, its *acceleration* with respect to absolute space is detectable. For when an object rotates then it is by definition accelerating, even if the rate of rotation is constant. This is because in physics, acceleration is defined as the rate of change of velocity, and velocity is speed *in a given direction*. Since rotating objects are constantly changing their direction of motion, their velocity is not constant, hence they are accelerating. The water's curved surface is just one example of what are called 'inertial effects'—effects produced by accelerated motion. Another example is the feeling of being pushed to the back of your seat when an aeroplane takes off. The only possible explanation of inertial effects, Newton believed, is the acceleration of the object experiencing those effects with respect to absolute space. For in a universe containing only the accelerating object, absolute space is the only thing that the acceleration could be relative to.

Newton's argument is powerful but not conclusive. For how does Newton know that the water's surface *would* curve upwards, if the rotating bucket experiment were done in a universe containing no other material objects? Newton simply assumes that the inertial effects we find in this world would remain the same in a world bereft of any other matter.

This is obviously quite a substantial assumption, and many people have questioned Newton's entitlement to it. So Newton's argument does not prove the existence of absolute space. Rather, it lays down a challenge to the defender of Leibniz to provide an alternative explanation of inertial effects.

Leibniz also faces the challenge of explaining the difference between absolute and relative motion without invoking absolute space. On this problem, Leibniz wrote that a body is in true or absolute motion 'when the immediate cause of the change is in the body itself'. Recall the case of the hang-glider and the observer on earth, both in motion relative to the other. To determine who is 'really' moving, Leibniz would say that we need to decide whether the immediate cause of the change (i.e. the relative motion) is in the hang-glider or the observer. This suggestion for how to distinguish absolute from relative motion avoids all reference to absolute space, but it is not very clear. Leibniz never properly explains what it *means* for the 'immediate cause of the change' to be in an object. But it may be that he intended to reject Newton's assumption that an object's motion, whether relative or absolute, can only be a fact about the object's relations to something else.

One of the intriguing things about the absolute/relational controversy is that it refuses to go away. Newton's account of space was intimately bound up with his physics, and Leibniz's views were a direct reaction to Newton's. So one might think that advances in physics since the 17th century would have resolved the issue. But this has not happened. Although it was once widely held that Einstein's theory of relativity had decided the issue in favour of Leibniz, this view has increasingly come under attack in recent years. More than 300 years after the original Newton/Leibniz exchange, the debate continues.

What are biological species?

Scientists often wish to classify the objects they are studying into general kinds or types. Geologists classify rocks as igneous, sedimentary, or metamorphic, depending on how they were formed. Chemists classify elements as metals, metalloids, or non-metals, depending on their physical and chemical attributes. The main function of classification is to convey information. If a chemist tells you that something is a metal, this tells you a lot about its likely behaviour. Classification raises some interesting philosophical issues. Mostly, these stem from the fact that a given set of objects can in principle be classified in many different ways. So how should we choose between them? Is there a 'correct' way to classify, or are all classification schemes ultimately arbitrary? These questions take on a particular urgency in the context of *biological* classification or taxonomy, which will be our concern here.

The basic unit of biological classification is the species. In traditional taxonomy each organism is first assigned to a species, denoted by a Latin name with two parts called a binomial. Thus you belong to *Homo sapiens*, your pet cat to *Felis catus*, and the mouse in your larder to *Mus musculus*. Species are then arranged into 'higher taxa'—genus, family, order, class, phylum, and kingdom—in a hierarchical fashion. Thus *Homo sapiens* is in the Homo genus, which is in the Hominid family, the Primate order, the Mammalian class, the Chordate phylum, and the Animal kingdom. This system of classification is called the *Linnean system*, named after the 18th-century Swedish naturalist Carl Linnaeus (1707–78) who invented it, and is still widely used today.

Our focus here will be on the first stage of the taxonomist's task, namely how to assign organisms to species. This is less straightforward than it may seem, primarily because biologists do not agree on what a species actually is, nor therefore on what criteria should be used for identifying species. Indeed competing definitions of a biological species, or 'species concepts' as these definitions are known, abound in modern biology. This lack of consensus is sometimes called 'the species problem'.

You may be surprised to hear that there is a species problem. From a layperson's perspective, there seems nothing very problematic about assigning organisms to species. After all, it is clear from even casual observation that living organisms are not all alike. Some are enormous while others are tiny; some move while others don't; some live for years and others for just a few hours. It is equally clear that this diversity is not continuous but clustered. Organisms appear to fall into a discrete number of types or kinds, many of which can be recognized by young children. A 3-year-old can confidently judge that two animals in the park are both dogs, even if they are of different breeds; and a biologist will confirm that the child is correct—the animals do indeed belong to the same species, namely *Canis familiaris*. Thus it is natural to think that there are objective divisions between living organisms that it is the job of the biologist to discover. On this view, species boundaries are 'out there' in the world awaiting discovery, rather than being imposed on the world by biologists. Most non-biologists appear to accept this view without question.

This commonsense viewpoint dovetails with the philosophical doctrine of 'natural kinds', variants of which have been popular since Aristotle. This doctrine holds that there are ways of grouping objects into kinds that are natural in the sense of corresponding to divisions that really exist in the world, rather than reflecting human interests. Chemical elements and compounds are paradigm natural kinds. Consider for example all the samples of pure gold in the universe. These samples are of the kind 'gold' because they are alike in a fundamental respect: their constituent atoms have atomic number 79. By contrast, a sample of fool's gold (iron pyrite) does not belong in that kind, despite being similar to gold in some respects, for it is a compound formed by atoms of a different sort (iron and sulphur). Philosophers who embrace scientific realism often argue that part of the job of any science is to discover the natural kinds in its domain.

The idea that species are the natural kinds of biology is an attractive one, but faces a number of challenges. One is that there may be an element of arbitrariness in what counts as a species. To see why, note that biologists often subdivide species into groupings such as breeds, varieties, and sub-species. For example the golden eagle *Aquila chrysaetos* is usually divided into six sub-species, which include the European, American, and Japanese golden eagle. The motivation for introducing sub-specific groupings is that some populations are recognizably different from each other, but not so different as to count as separate species. But how do we draw a sharp line? Darwin offers an interesting discussion of this point in *The Origin of Species*, arguing that no clear demarcation exists between species, sub-species, and varieties. He concluded: 'it will be seen that I look at the term species as one arbitrarily given, for the sake of convenience, to a set of individuals closely resembling each other, and that it does not essentially differ from the term variety, which is given to less distinct and more fluctuating forms.'

Darwin's suggestion that whether a set of individuals counts as a species is arbitrary is surprising; certainly it does not fit with the idea of species as natural kinds. But is Darwin right? In the 20th century, many evolutionary biologists became convinced that species are in fact real units in nature, not arbitrary groupings, on the grounds that they are *reproductively isolated*. This means that the organisms within a species can interbreed with each other but not with those of other species. Defining biological species in terms of reproductive isolation was championed by the German biologist Ernst Mayr, and became known as the 'biological species concept' (BSC). Proponents of the BSC reject Darwin's argument that the variety/species distinction is arbitrary. On their view, the European and American golden eagles are varieties, not separate species, since they can in principle interbreed and produce viable offspring (even if they do so rarely). By contrast, the spotted eagle and the golden eagle are separate species, since their members cannot interbreed.

The BSC is widely used in contemporary biology, but it has limitations. It only applies to sexually reproducing organisms; however, many living organisms reproduce asexually, including most single-celled organisms, some plants and fungi, and a few animals. So the BSC is at best a partial solution to the species problem. Moreover, reproductive isolation is not always a hard and fast matter. Closely related species who live in adjacent locations often have 'hybrid zones' where their ranges meet; in these zones, a limited amount of hybridization takes place, in which fertile offspring are produced at least some of the time, but the two species retain their distinct identities. Hybrid zones often arise when one species is in the process of splitting into two. Among plants, in particular, hybridization between organisms that belong to clearly distinct species is quite common.

Still more problematic for the BSC is the phenomenon of ring species. This occurs when a species comprises a number of local populations arranged geographically in a ring; each population can interbreed with an immediate neighbour, but the populations at the end of the chain cannot. For example, suppose population A can interbreed with B, B with C, C with D, D with E, but A and E cannot interbreed (see Figure 8). Ring species constitute a kind of paradox for the BSC. To see why, ask yourself whether A and E belong to the same species or not. Since they cannot interbreed, the answer should be 'no'. However A and B are con-specific by the interbreeding criterion, as are B and C, C and D, and D and E, so surely A and E *must* be con-specific? It is quite unclear what to say about this situation, on the BSC. So Darwin's insistence that there is an element of arbitrariness in what gets counted as a species is not entirely defused by the focus on reproductive isolation.

The underlying source of the species problem is evolution itself. Modern biology teaches us, following Darwin, that all living organisms derive from a common ancestor. However traditional Linnean taxonomy derives from a time when creationism was the

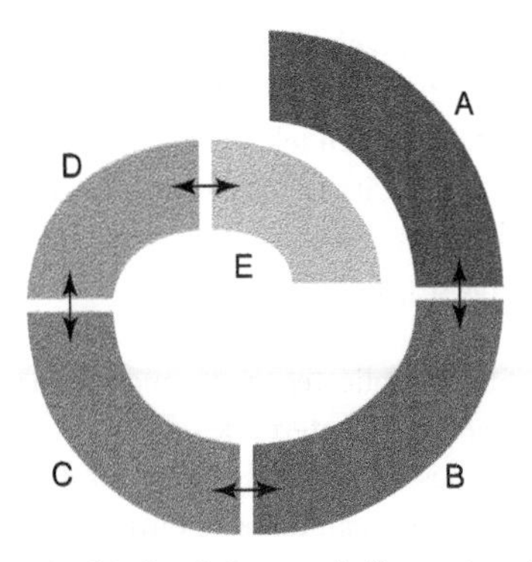

8. Ring species. A double-headed arrow indicates interbreeding.

dominant worldview. On a creationist view, in which God created each species separately, it is natural to expect that all organisms can be assigned to species unambiguously. But on an evolutionary view there is no reason to expect this. For evolutionary change is gradual—the process by which an ancestral species gives rise to a daughter species typically takes many thousands of years. Often it involves one species gradually splitting into two, with reproductive links between the two daughter species eventually being severed. So transitional forms, and populations whose status as species is unclear, are only to be expected. Moreover, there is no reason to expect that a single species definition will work for all organisms, from bacteria to multi-celled animals.

Evolution also teaches us that *variation* among organisms is likely to be pervasive. For variation is the engine that drives natural selection: if the organisms in a species do not vary then natural selection cannot operate. The significance of this is that it undermines the commonsense idea that the members of a biological species must all possess some essential feature, e.g. some genetic property, which sets them apart from non-members. This idea is part of the 'natural kind' view of species, and is something that many non-biologists appear to believe. Empirically, however, there is extensive genetic variation among the individuals within a typical species, which sometimes exceeds the genetic variation between closely related species. This is not to deny that biologists can often tell what species an organism belongs to by sequencing its DNA. However this is not always possible, and it does not show that membership of a species is determined by a fixed 'genetic essence'.

Evolution therefore complicates the taxonomic enterprise considerably. However the enterprise must go on, for dividing organisms into species is practically indispensable. If an ornithologist comes across an unusual bird, for example, the first thing they will want to know is what species it is from, as this provides valuable information about its traits, behaviour, and ecology. The situation was eloquently described by the English biologist John Maynard Smith, who wrote: 'any attempt to divide all living organisms, past and present, into sharply defined groups between which no intermediates exist, is foredoomed to failure. The taxonomist is faced with a contradiction between the practical necessity and the theoretical impossibility of his task.' So in practice biologists continue to treat species as if they were sharply defined kinds, in the knowledge that this is only an approximation to reality.

From the late 1960s onwards, evolutionary biology has increasingly converged on the idea that classification should be done in a way that is 'consistent' with evolution. This was the *leitmotif* of the movement known as 'phylogenetic systematics', founded by the German entomologist Willi Hennig. The key idea was to recognize as genuine only those biological groupings that are *monophyletic*, which means that they contain all and only the descendants of some common ancestor. Many traditional taxonomic groupings, e.g. the class *Reptilia* (the reptiles), have turned out not to be monophyletic—in this case, because the ancestor of all the reptiles also gave rise to the birds (see Figure 9). So proponents of phylogenetic classification insist that *Reptilia* is not a genuine taxon, and has no place in a correct taxonomy. Phylogenetic

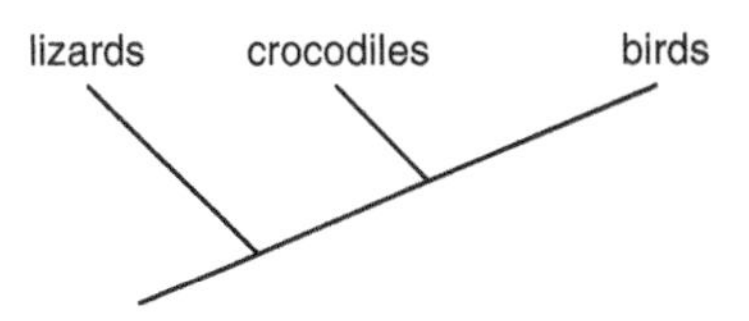

9. Phylogenetic relationship between lizards, crocodiles, and birds.

systematics is mainly concerned with how to delimit higher taxa, rather than species. However the monophyly criterion can be applied to individual species, yielding what is known as the 'phylogenetic species concept'; in effect, this concept is an attempt to formulate precisely the intuitive idea that the organisms within a species should be more closely related to each another than to the members of other species.

From a philosophical viewpoint, the significance of the phylogenetic approach is its implication that two organisms belong to the same group—species or higher taxon—because of their shared ancestry, rather than their intrinsic similarity. A thought experiment can help flesh this out. Suppose that scientists find an organism on Mars that is not derived from earthly biological matter, but nonetheless is completely indistinguishable from a common housefly. (This is incredibly unlikely, of course, but it is logically conceivable.) The Martian specimen looks like a housefly, can interbreed with houseflies back on earth, and cannot be told apart from true houseflies by any genetic tests. Is it a housefly? If species were natural kinds, the answer would presumably be yes. But on a phylogenetic view, the answer is no. To be of the housefly species (*Musca domestica*), an organism needs to have the appropriate pattern of ancestry, irrespective of what intrinsic features it has.

This tallies with an intriguing suggestion made by the biologist Michael Ghiselin and the philosopher David Hull in the 1970s. They argued that a biological species should not be regarded as a kind or type at all, which is the traditional assumption, but rather as a complex individual extended in space and time. Like an individual organism, a species comes into existence at a particular place and time, has a finite lifespan, and then goes extinct. By contrast, a genuine kind is spatiotemporally unrestricted. Consider gold. A piece of matter anywhere in the universe counts as gold, irrespective of its origin, so long as it has atomic number 79. So in principle, all the gold in the universe could be destroyed and then years later some more could be synthesized. But species are not like this, Ghiselin and Hull argued. Once a species goes extinct it can never come back into existence as a matter of logic, any more than you or I can survive our deaths.

The idea that species are individuals sounds strange at first but makes sense on reflection. Certainly, species are unlike 'ordinary' individuals in that their constituent parts, i.e. the organisms they contain, are not joined together. However this difference is fairly superficial. The insects in an ant colony are not joined together either, but we are happy to regard the colony as an individual thing. Treating species as individuals has distinct advantages. One is that it fits well with the principles of phylogenetic systematics. Another is that it helps reconcile the widespread intuition that species are 'real' units in nature, not arbitrary groupings, with the fact that intra-specific genetic variation is widespread and that species do not have 'genetic essences'. These facts make good sense if we view species as complex individuals, but make much less sense if we regard them as natural kinds.

Is the mind modular?

It is a striking fact that humans are able to perform a diverse array of cognitive tasks, often with fairly little conscious effort. By 'cognitive tasks' we do not just mean things like solving crossword puzzles, but also everyday tasks like crossing the road safely, catching a ball, understanding what other people say, recognizing other people's faces, and more. Such tasks are so familiar to us that they are easily taken for granted; however our ability to perform them is really quite remarkable. No robot comes close to being able to perform most of these tasks as well as an average human, despite considerable expense. Somehow or other, our brains enable us to perform complex cognitive tasks with minimal effort. Explaining how this could be is an important part of the discipline known as cognitive psychology.

Our focus is an ongoing debate among cognitive psychologists concerning the architecture of the human mind. According to one view, the human mind is a 'general-purpose problem solver'. This means that the mind contains a set of general problem-solving skills, or 'general intelligence', which it applies to an indefinitely large number of different tasks. So one and the same set of cognitive capacities are employed, whether a person is trying to count marbles, decide which restaurant to eat in, or learn a foreign language—these tasks represent different applications of the person's general intelligence. According to a rival view, the human mind contains a number of specialized sub-systems or modules, each of which is designed for performing a specific task and cannot do anything else. This is known as the *modularity of mind* hypothesis.

One example of modularity comes from the work of the linguist Noam Chomsky on language acquisition in the 1960s. Chomsky insisted that a child does not learn language by overhearing adult conversation and then using their 'general intelligence' to figure out the rules of the language being spoken. This is impossible, he argued, for the linguistic data that children are exposed to are too limited, and vary greatly from child to child; yet all children acquire language by the same age. Chomsky argued that there is a distinct module called the 'language acquisition device' in every child's brain. The device operates automatically, and its sole function is to enable the child to acquire language. It does this by encoding the principles of 'universal grammar' that all human languages obey, enabling the child to learn the grammar of any language given appropriate prompting. Chomsky provided an array of impressive evidence for the claim that language acquisition is a modular capacity—such as the fact that even those with low 'general intelligence' can usually learn to speak perfectly well.

Some of the most compelling evidence for the modularity hypothesis comes from studies of patients with brain damage, known as 'deficit studies'. If the human mind is a general-purpose problem solver, we would expect damage to the brain to affect all cognitive capacities more or less equally. But this is not what we find. On the contrary, brain damage often impairs some cognitive capacities but leaves others untouched. For example, damage to a part of the brain known as Wernicke's area leaves patients unable to understand speech, though they are still able to produce fluent, grammatical sentences. This strongly suggests that there are separate modules for sentence production and comprehension—for this would explain why loss of the latter capacity does not entail loss of the former. Other brain-damaged patients lose their long-term memory (amnesia), but their short-term memory and ability to speak and understand are unimpaired. Again, this seems to speak in favour of modularity and against the view of the mind as a general-purpose problem solver.

Though compelling, neuropsychological evidence of this sort does not settle the modularity issue decisively. Such evidence is relatively sparse—we obviously cannot damage people's brains at will just to see how their cognitive capacities are affected. In addition, there are disagreements about how the data should be interpreted, as is usual in science. Some argue that the observed pattern of cognitive impairment in brain-damaged patients does not imply modularity. Even if the mind were a general-purpose problem solver, i.e. non-modular, it is still possible that distinct cognitive capacities might be differentially affected by brain damage, they argue. So we cannot simply read off the architecture of the mind from deficit studies; at best, the latter provide fallible evidence for the former.

Much of the recent interest in modularity is due to the work of Jerry Fodor, an American philosopher and psychologist. In a 1983 book entitled *The Modularity of Mind* Fodor offered a novel account of what a module is, and some interesting ideas about which cognitive capacities are modular and which not. Fodor argued that mental modules have a number of distinguishing features, of which three are particularly important: (i) they are *domain-specific*, (ii) their operation is *mandatory*, and (iii) they are *informationally encapsulated*. Non-modular systems lack these features. Fodor then argued that the mind is partly though not wholly modular: we solve some cognitive tasks using specialized modules, others using our general intelligence.

To say that a cognitive system is domain-specific is to say that it is specialized: it performs a limited, precisely circumscribed set of tasks. Chomsky's postulated language acquisition device is an example. The sole function of this device is to enable the child to learn language—it doesn't help the child learn to play chess, or to count, or anything else. So the device ignores non-linguistic inputs. To say that a cognitive system is mandatory is to say that we cannot choose whether to put it into operation. The perception of language is an example. If you hear a sentence uttered in a language you know, you cannot help but hear it as the utterance of a sentence. If someone asked you to hear the sentence as 'pure noise' you couldn't obey them however hard you tried. Fodor points out that not all cognitive processes are mandatory in this way. Thinking clearly is not. If someone asked you to think of the scariest moment in your life, or of what you would do if you won the lottery, you clearly could obey their instructions. So thinking and language perception are quite different in this regard.

What about informational encapsulation? This notion is best illustrated by an example. Look at the two parallel lines in Figure 10. To most people, the top line looks slightly longer than the bottom one. But in fact this is an optical illusion, known as the Müller-Lyer illusion. The lines are actually equal in length.

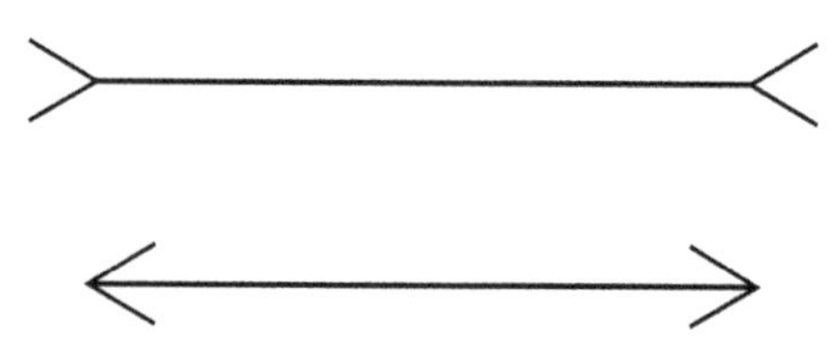

10. The Müller-Lyer illusion.

Various explanations have been suggested for why the top line looks longer, but they need not concern us here. The point is this: the lines continue to look unequal in length *even when you know it's an optical illusion.* According to Fodor, this simple fact has important implications for understanding the architecture of the mind. For it shows that the information that the two lines are equal in length is stored in a region of the cognitive mind to which our perceptual mechanisms do not have access. If visual perception was not encapsulated in this way, but could make use of all the information stored in the mind, then the illusion would disappear as soon as you were told that the lines were actually equal in length.

Another possible example of information encapsulation comes from the phenomenon of human phobias. Consider ophidiophobia, or fear of snakes, which is widespread among humans. Even if you know that a particular snake isn't dangerous, e.g. because you've been told that its poison glands have been removed, you are still quite likely to be scared of it and will not want to touch it. This sort of phobia can often be overcome by training, but that is a different matter. The key point is that the information that the snake isn't dangerous is inaccessible to the part of your mind which produces in you the reaction of fear when you see a snake. This suggests that there may be an inbuilt, informationally encapsulated 'fear of snakes' module in every human being.

You may wonder why the modularity of mind issue is a matter for philosophers. Surely it is just a question of scientific fact whether the mind is modular or not? In a way this is true, but the debate has a number of philosophical dimensions. One concerns how to count cognitive tasks and modules. Advocates of modularity hold that the mind contains specialized modules for performing different cognitive tasks; their opponents deny this.

But how do we decide whether two cognitive tasks are the same or different? Is facial recognition a single cognitive task or does it comprise two distinct cognitive tasks: recognizing male faces and recognizing female faces? Are long division and multiplication different cognitive tasks, or are they both part of the more general task of doing arithmetic? Questions of this sort are conceptual or philosophical, rather than straightforwardly empirical, and are potentially crucial to the modularity debate. For suppose an opponent of modularity produces experimental evidence to show that we use a single cognitive capacity to perform many different types of cognitive task. Their opponent might accept the experimental data, but argue that the cognitive tasks in questions are all of the *same* type, and thus that the data are perfectly compatible with modularity. So in order for the modularity debate to be well defined, a principled way of counting cognitive tasks and modules is needed.

A second philosophical dimension arises because both proponents and opponents of modularity have employed a priori arguments, in addition to direct empirical evidence, in support of their view. Fodor himself argued that while perception and language are probably modular, thinking and reasoning cannot be since they are 'holistic'. To understand Fodor's argument, suppose you are a member of a jury deliberating over what verdict to return. How will you go about your task? One thing you will consider is whether the defendant's story is logically consistent—is it free from contradiction? And you will ask yourself how strong the evidence against the defendant really is. The reasoning skills you apply here—testing for logical consistency and assessing evidence—are general; they are not specifically designed for use in jury service. So the cognitive capacities you bring to bear in deliberating the defendant's guilt are not domain-specific. Nor is their operation mandatory—you have to consciously consider whether the defendant is guilty, and can stop doing so whenever you want to, e.g. during the lunch break. Finally, there is no information encapsulation either. Your task is to decide whether the defendant is guilty *all things considered*, so you may have to draw on any of the background information that you possess, if you consider it relevant. For example, if the defendant twitched nervously under cross-examination and you believe that nervous twitching is often a sign of guilt, you will probably draw on this belief in reaching your verdict. So there is no store of information which is inaccessible to the cognitive mechanisms you employ to reach your verdict (though the judge may tell you to ignore certain things). In short, there is no module for deciding whether a defendant is guilty. You tackle this cognitive problem using your general intelligence.

Some psychologists have gone further than Fodor and suggested that the mind is entirely modular; this is known as the 'massive modularity hypothesis'. Proponents of massive modularity argue on general grounds that we should expect the human mind to exhibit a modular organization. Their main argument comes from considerations of Darwinian adaptiveness. The human mind is widely supposed to have evolved in the Pleistocene epoch, to allow our hominid ancestors to solve the social and environmental challenges they faced. The suggestion is that a modular organization provides the most efficient way of solving these challenges and so behaving adaptively. A set of dedicated modules, each specialized for a given task, allows faster and more accurate problem solving. The Swiss army knife analogy is relevant here. Clearly, having separate tools for can-opening, bottle-opening, and screw-driving is preferable to having a single all-purpose tool that can do each of these things. Similarly, a mind with separate modules for face recognition, language acquisition, and mate choice, for example, will be more efficient than one that is a general-purpose problem solver. So considerations of optimal design tell in favour of mental modularity, the argument goes.

A related argument holds that a human cannot possibly acquire all the information needed to behave adaptively in its own lifetime. The opportunities for learning are simply too few, it is argued. Therefore the mind needs to contain innate information, encapsulated in a module, which will allow the child to develop the appropriate cognitive skills and thus behave adaptively. Chomsky's language acquisition module illustrates this point. The module contains innate information about the grammar of every human language, enabling the child to acquire a language given quite minimal input. This 'gap' between the information needed to solve a cognitive task and the information that can be acquired by learning is often used to argue in favour of a modular cognitive organization. Strictly speaking, though, it is an argument for the mind's containing innate information, rather than for modularity per se. These are logically distinct ideas, but in practice defenders of modularity tend to believe in innate information and vice versa.

This connection between modularity and innateness points to another respect in which the modularity debate is philosophically significant. The idea that the mind contains innate information is sharply at odds with traditional empiricist philosophy, according to which all knowledge comes from experience. In the 17th and 18th centuries, empiricists such as John Locke and David Hume argued that at birth the human mind is a 'tabula rasa', or blank slate, with nothing written on it. It is only through experience that a human comes to have concepts and knowledge. This empiricist doctrine is a venerable one, and at first sight strikes many as obviously true. But Chomsky argued that his language acquisition module, containing as it does innate information about universal grammar, directly disproves this aspect of the empiricist philosophy. If Chomsky's argument is correct—a hotly debated point—it provides an interesting example of how scientific findings can inform traditional philosophical debates.

It is too early to say whether the massive modularity thesis will prove tenable. The a priori arguments (for and against) cannot themselves settle the issue; direct evidence is needed. Fodor himself rejects massive modularity, and as a result is pessimistic about the possibility of cognitive psychology ever explaining the workings of the human mind. He believes that only modular systems can be studied scientifically—non-modular systems, because they are not informationally encapsulated, are much more difficult to model. So according to Fodor the best research strategy for cognitive psychologists is to focus on perception and language, leaving aside thinking and reasoning. But unsurprisingly, this aspect of Fodor's thought is highly controversial.

Chapter 7
Science and its critics

Many people take it for granted that science is a good thing, for obvious reasons. After all, science has given us electricity, safe drinking water, penicillin, and air travel—all of which have undoubtedly benefited humanity. But despite these impressive contributions to human welfare, science is not without its critics. Some argue that society spends too much money on science at the expense of the arts; others observe that science has given us technological capabilities we would be better off without, e.g. weapons of mass destruction. Some feminists argue that science is inherently male-biased; those of religious persuasion often feel that science threatens their faith; and anthropologists have accused Western science of arrogantly assuming its superiority over the knowledge and beliefs of indigenous cultures. This does not exhaust the list of criticisms to which science has been subject, but here we confine our attention to three that are of particular philosophical interest.

Scientism

The word 'scientific' has acquired a peculiar cachet in modern times. If someone accuses you of behaving unscientifically, they are almost certainly criticizing you. Scientific conduct is rational and praiseworthy; unscientific conduct is irrational and worthy of contempt. It is difficult to know why calling something scientific should carry these connotations, but it probably relates to the high status which science holds in modern society. Scientists are treated as experts, their opinions regularly sought on matters of social importance. Of course, everybody recognizes that scientists sometimes get it wrong; for example, scientific advisers to the British government in the early 1990s declared that 'mad cow' disease posed no threat to humans, only to be proved tragically mistaken. But occasional hiccups of this sort tend not to shake the public faith in science, nor the esteem in which scientists are held. In many countries, scientists are viewed much as religious leaders used to be: possessors of specialized knowledge that is inaccessible to the laity.

'Scientism' is a pejorative term used by some philosophers to describe what they see as science worship, or an over-reverential attitude towards modern science. Opponents of scientism argue that science is not the only valid form of intellectual endeavour, and not the only way of understanding the world. They often stress that they are not anti-science per se; what they are opposed to is the assumption that scientific methods are necessarily applicable to every subject matter. So their aim is not to attack science but to put it in place, by rejecting the idea that scientific knowledge is all the knowledge there is.

Scientism is a rather vague doctrine, and, given that the term has a pejorative usage, few would explicitly admit to believing it. Nonetheless, something quite like science worship is a genuine feature of the intellectual landscape. This is not necessarily a bad thing—perhaps science deserves to be worshipped. But it is certainly a real phenomenon. One field which is often accused of science worship is contemporary anglophone philosophy (of which philosophy of science is just one branch). Traditionally philosophy is regarded as a humanities subject, despite its close historical links to mathematics and science, and with good reason. For philosophy asks questions about the nature of knowledge, morality, and human well-being, for example, which do not seem soluble by scientific methods. No branch of science tells us how we should lead our lives, what knowledge is, or what human flourishing involves; these are quintessentially philosophical questions.

In the light of this, it may seem surprising that some philosophers insist that science is the only legitimate path to knowledge. Questions that cannot be resolved by scientific means are not genuine questions at all, they hold. This view was endorsed by the famous 20th-century English philosopher Bertrand Russell, who wrote: 'whatever knowledge is attainable, must be attained by scientific methods; and what science cannot discover, mankind cannot know'. The grounds for the view lie in a doctrine called 'naturalism', which stresses that we human beings are part and parcel of the natural world, not something apart from it, as was once believed. Since science studies the whole of the natural world, surely it should be capable of revealing the complete truth about the human condition, leaving nothing left for philosophy? On this view, philosophy has no distinctive subject matter of its own. Insofar as it serves a useful role at all, it involves 'clarifying scientific concepts'—clearing the brush so that scientists can get on with their work.

Not surprisingly, many philosophers reject this subordination of their discipline to science; this is one source of opposition to scientism. They argue that philosophical enquiry has its own proprietary methods, which can reveal truths of a sort that science cannot. Proponents of this view allow that philosophy should aim to be *consistent* with the sciences, in the sense of not advancing claims which conflict with what science teaches us. And they typically accept that we humans are part of the natural order, so not exempt from the scope of science. But it does not follow, they argue, that science is the only legitimate source of knowledge about the world.

What exactly are these methods of philosophical enquiry? They include logical reasoning, the use of thought experiments, and what is called 'conceptual analysis', which tries to delimit a particular concept by relying on our intuitions about whether a particular case falls under it. For example, a classical philosophical question asks whether knowledge is identical to true belief. Most philosophers say that the answer is 'no', on the grounds that we can construct cases in which a person does truly believe a particular proposition but cannot be said to know it. (For example, suppose you believe that it is ten past six because that is what your watch says; in fact your watch is broken, but by chance the time *is* actually ten past six! Your belief is therefore true, but intuitively you did not *know* that it was ten past six as you simply 'got lucky'.) So by using conceptual analysis, we can establish that knowledge and true belief are not identical—which is a substantive philosophical truth. This is just one example; but it illustrates the idea that philosophical enquiry can yield genuine knowledge using its own non-scientific methods.

How should this debate be assessed? On the one hand, there are certainly examples of philosophical questions which appear to be genuine, to lie outside the provenance of any science, and to be answerable by the distinctive methods of philosophers. However, against this, many of the questions discussed in the history of philosophy, e.g. about perception, imagination, and memory, have turned out to be matters for the empirical sciences, in particular psychology. Indeed the pool of questions classed as 'philosophical' has shrunk over the centuries, as more and more get appropriated by science. Moreover, the idea that philosophical enquiry and scientific enquiry are autonomous, each relying on their own methods, has been criticized as wishful thinking; opponents point out that while there is certainly progress in science, progress in philosophy is rather harder to discern.

An analogous issue concerns the relation between the natural and social sciences. Just as philosophers complain of 'science worship' in their discipline, so social scientists complain of 'natural science worship' in theirs. It is often felt that natural sciences such as physics, chemistry, and biology are in a more advanced state than social sciences such as economics, sociology, and anthropology; the former can formulate precise laws with great predictive power, while the latter usually cannot. Why should this be so? It can hardly be because natural scientists are smarter than social scientists. One possible answer is that the *methods* of the natural sciences are superior. If this is correct, then what the social sciences need to do to catch up is to ape the methods of the natural sciences. To some extent this has already happened. The increasing use of mathematics in the social sciences may be partly a result of this attitude. Physics made a great leap forward when Galileo took the step of applying mathematical language to the description of motion; so it is tempting to think that a comparable leap forward might be achievable in the social sciences if a comparable way of 'mathematicizing' their subject matter can be found.

However, some social scientists resist the suggestion that they should look up to the natural sciences in this way, arguing that the methods of natural science are not necessarily appropriate for studying social phenomena. They typically deny that the social sciences are impoverished vis-à-vis the natural sciences, pointing out that the complexity of social phenomena, and the fact that controlled experiments usually cannot be done, mean that finding precise laws with predictive power is not an appropriate benchmark of success.

An influential version of this argument derives from the 19th-century German sociologists Wilhelm Dilthey and Max Weber. They argued that social phenomena must be understood from the viewpoint of the actor(s) responsible for them. What distinguishes social from natural phenomena is that the former are the result of the intentional action of humans. Thus a special type of understanding, called *verstehen*, is needed for social scientific enquiry; this involves trying to grasp the subjective meaning that a social action has for the actor. For example, an anthropologist studying a religious ritual needs to understand the significance that the ritual has for the participants; a purely 'objective' analysis, of the sort that could be had by applying the methods of natural science, cannot yield a genuine understanding of the ritual, since it neglects the crucial matter of the ritual's meaning. The doctrine of *verstehen* thus posits a sharp discontinuity between the natural and social sciences. The doctrine had a significant influence on the development of anthropology and sociology, in particular, in the 20th century.

Neither the scientism issue nor the parallel issue about natural and social science is easy to resolve. In part, this is because it is not fully clear what exactly the 'methods of science', or the 'methods of natural science', actually comprise—a point that is often overlooked by both sides in the debate. If we want to know whether the methods of science are applicable to every subject matter, or whether they are capable of answering every important question, we obviously need to know what exactly those methods *are*. But as we have seen, this is less straightforward a question than it seems. Certainly we know some of the main features of scientific enquiry: experimental testing, observation, theory construction, and inductive inference. But this list does not provide a precise definition of 'the scientific method'. Nor is it obvious that such a definition could be provided. Science changes rapidly over time, so the assumption that there is a fixed, unchanging 'scientific method', used by all scientific disciplines at all times, is not inevitable. But this assumption is implicit both in the claim that science is the only route to knowledge *and* in the counter-claim that some questions cannot be answered by scientific methods. This suggests that, to some extent at least, the debate about scientism may rest on a false presupposition.

Science and religion

The tension between science and religion is old. Perhaps the best-known example is Galileo's clash with the Catholic Church.

In 1633 the Inquisition forced Galileo to publicly recant his Copernican views and condemned him to spend his last years under house arrest in Florence. The Church objected to the Copernican theory because it contravened the holy Scriptures, of course. In recent years, the most prominent science/religion clash has been the bitter dispute between Darwinists and proponents of 'intelligent design' in the United States, which will be our focus here.

Theological opposition to Darwin's theory of evolution is nothing new. When *The Origin of Species* was published in 1859, it immediately attracted criticism from churchmen in England. The reason is obvious: Darwin's theory maintains that all current species, including humans, have descended from common ancestors over a long period of time. This theory clearly contradicts the Book of Genesis, which says that God created all living creatures over a period of six days. Some Darwinians have tried to reconcile their belief in evolution with their Christian faith by arguing that the Book of Genesis should not be interpreted literally—it should be regarded as allegorical, or symbolic. However, in the USA, many evangelical Protestants have been unwilling to bend their religious beliefs to fit scientific findings. They insist that the biblical account of creation is literally true, and that Darwin's theory of evolution is therefore completely wrong.

This opinion is known as 'creationism', and is accepted by some 40 per cent of the adult population in the USA. Creationism is a powerful political force, and over the years has had considerable influence on the teaching of biology in American high schools, to the dismay of scientists. Since the American constitution prohibits the teaching of religion in public schools, 'creation science' was invented—which claims that the biblical account of creation is a better scientific explanation of life on earth than Darwin's theory of evolution. So teaching biblical creation does not violate the constitutional ban, for it counts as science, not religion! In 1981 a law was passed in Arkansas calling for biology teachers to give 'equal time' to evolution and creation science. However this was overruled by a federal judge the following year, and in 1987 a Supreme Court judgment ruled that teaching creation science in public schools was unconstitutional.

Following these legal defeats, the creation science movement cleverly rebranded itself under the label 'intelligent design'. The name is an allusion to an old argument for the existence of God, known as the 'argument from design', which says that the existence of complex biological organisms can only be explained by supposing that an intelligent deity created them; this deity is usually identified with the Christian God. The argument from design was part of the intellectual mainstream in the pre-Darwinian era, but is of course rejected by contemporary biologists. Proponents of intelligent design have resuscitated the argument, claiming that biological organisms exhibit 'irreducible complexity' which could not have evolved by Darwinian means, and is thus proof of God's handiwork. An 'irreducibly complex' system is one with a number of interacting parts each of which is critical to the system's functioning—remove or alter any one of the parts and the system breaks down. It is true that biological organisms, and indeed individual cells, are complex in this sense, as their functioning depends on the coordinated activity of many biochemical components. This sort of interdependence could not have evolved by natural selection, claim the intelligent design camp.

Despite its recent prominence, this argument is old wine in new bottles. In *The Origin of Species*, Darwin himself wondered how the vertebrate eye, a highly complex organ, might have evolved by natural selection, noting that at first blush it seems 'absurd'. However, Darwin believed that the difficulty could be resolved by imagining a sequence leading from a simple eye (perhaps just a few light-sensitive cells) to modern eyes by a gradual series of incremental improvements, each of which conferred a selective advantage. In this way an organ of great complexity, with finely tuned components, could have evolved by natural selection. Darwin himself could only guess as to what the intermediate stages of eye evolution were. But recent scientific work has offered detailed insight into the probable sequence of stages, based on studying the embryonic development of the eye, and performing detailed genetic analyses, across vertebrate species. So the suggestion that the eye could not have arisen by natural selection has been successfully rebutted. The moral generalizes: there is no evidence to support the idea that organisms exhibit any features that could not have resulted from an evolutionary process.

In addition to their emphasis on 'irreducible complexity', proponents of intelligent design have tried to undermine the Darwinian worldview in other ways. They argue that the evidence for Darwinism is inconclusive, so Darwinism should not be regarded as established fact but rather as 'just a theory'. In addition, they have focused on various internal disputes among Darwinians, and picked on a few incautious remarks by individual biologists, in an attempt to show that disagreeing with the theory of evolution is scientifically respectable. They conclude that since Darwinism is 'just a theory', students should be exposed to alternative theories too—such as the theory that an intelligent deity created all living organisms.

In a way, it is quite correct that Darwinism is 'just a theory' and not proven fact. As we saw in Chapter 2, it is never possible to *prove* that a scientific theory is true, in the strict sense of proof, for the inference from data to theory is invariably non-deductive. But this is a general point—it has nothing to do with the theory of evolution per se. By the same token, we could argue that it is 'just a theory' that the earth goes round the sun, or that water is made of H_2O, or that unsupported objects tend to fall, so students should be presented with alternatives to each of these. But proponents of intelligent design do not argue this. They are not sceptical about science as a whole, but about the theory of evolution in particular. So if their position is to be defensible, it cannot simply turn on the point that our data doesn't guarantee the truth of Darwin's theory. For the same is true of every scientific theory, and indeed of most commonsense beliefs too.

Another intelligent design argument is that the fossil record is patchy, particularly when it comes to the supposed ancestors of *Homo sapiens*. There is some truth in this charge. Evolutionists have long puzzled over the gaps in the fossil record. One persistent puzzle is why there are so few 'transition fossils'—of creatures intermediate between two species. If later species evolved from earlier ones as Darwin's theory asserts, surely transition fossils should be common? However this is not a good argument against Darwin's theory. For fossils are not the only or even the main source of evidence for the theory of evolution. Other sources include comparative anatomy, embryology, biogeography, and genetics. Consider for example the fact that humans and chimpanzees share 98 per cent of their DNA. This and thousands of similar facts make perfect sense if the theory of evolution is true, and thus constitute excellent evidence for the theory. Of course, a proponent of intelligent design can also explain this fact, by saying that the designer chose to make humans and chimpanzees genetically similar for reasons of his (or her) own. But the possibility of giving 'explanations' of this sort simply shows that Darwin's theory is not logically entailed by the evidence, so in principle other explanations can be concocted. This methodological point is correct, but shows nothing special about Darwinism.

Though the arguments of the intelligent design camp are uniformly unsound, the controversy does raise serious questions concerning science education. How should the tension between science and faith be dealt with in a secular education system? Who should determine the content of high-school science classes? Should parents who don't want their children to be taught about evolution, or some other scientific matter, be overruled by the state? These questions normally receive little public attention, but the clash between Darwinism and intelligent design has brought them to the fore.

Is science value-free?

Everybody would agree that scientific knowledge has sometimes been used for unethical ends—to make nuclear and chemical weapons, for example. But such cases do not show that there is something ethically objectionable about scientific knowledge itself. It is the *use* to which that knowledge is put that is unethical. Indeed many philosophers would say that it makes no sense to talk about science or scientific knowledge being ethical or unethical per se. For science is concerned with facts, and facts in themselves have no ethical significance. It is what we do with those facts that is right or wrong, moral or immoral. On this view, science is essentially a *value-free* activity—its job is just to provide information about the world. What society chooses to do with that information is another matter.

Not all philosophers accept this picture of science as neutral with respect to questions of value, nor the underlying fact/value dichotomy on which it rests. Some claim that scientific enquiry is invariably laden with value judgements. One argument for this stems from the obvious fact that scientists have to choose what to study—not everything can be examined at once. So judgements about the relative importance of different possible objects of study have to be made, and these are value judgements, in a weak sense. Another argument stems from the fact that any set of data can in principle be explained in more than one way. A scientist's choice of theory will thus never be uniquely determined by their data. Some philosophers take this to show that values are invariably involved in theory choice, and thus that science cannot be value-free. A third argument is that scientific knowledge cannot be divorced from its intended applications in the way that value-freedom would require. On this view, it is naive to picture scientists as disinterestedly doing research for its own sake, without a thought for its practical applications. The fact that much scientific research today is funded by the private sector lends some credence to this view.

Though interesting, these arguments are all somewhat abstract—they seek to show that science could not be value-free as a matter of principle, rather than identifying actual cases of values playing a role in science. But specific allegations of value-ladenness have also been made. Here we focus on two examples, one from psychology/biology and the other from medicine.

Our first case concerns the discipline of evolutionary psychology, which tries to understand the psychological make-up of humans, and their resulting behaviour, by applying Darwinian principles. At first blush this project sounds perfectly reasonable. For humans are just another species of animal, and biologists agree that Darwinian theory can explain a lot about animal behaviour and its psychological underpinnings. For example, there is an obvious Darwinian explanation for why mice have an instinctive fear of cats. In the past, mice with this instinctive fear tended to leave more offspring than ones without, as the latter got eaten; assuming that the instinct was genetically based, and thus transmitted from parents to offspring, over many generations it would have spread through the population. Evolutionary psychologists believe that many aspects of human psychology can be given a Darwinian explanation of this sort.

To illustrate, consider human mating preferences. There is evidence that males and females systematically differ in the attributes they seek in their mating partners. (The strength of this evidence is a matter of debate.) A large cross-cultural survey by David Buss found that males on average preferred their female marriage partner to be younger than them, and to be of an age that coincides closely with peak female fertility (about 24 years). By contrast, females preferred to marry men who were older than them. Moreover, physical attractiveness mattered more to males, whereas earning potential mattered more to females. Buss and other evolutionary psychologists argue that these preferences have a Darwinian explanation. From an evolutionary viewpoint, the best strategy for a male is to find a female mate with high reproductive potential, as this maximizes the number of offspring he can have with her. Females should prefer to find a high-status male, who controls resources and is able to provide for the offspring. (This difference in optimal mating strategy stems from the fact that females have a limited supply of eggs, while males have effectively unlimited sperm, so offspring care matters more for females.) Therefore, it is argued, the mating preferences of modern humans can be explained by Darwinian natural selection.

Though the idea that humans' psychological traits have evolved by natural selection is plausible, evolutionary psychology is a controversial field, and its practitioners have been accused of ideological bias. The controversy dates back to the 'sociobiology wars' of the 1970s and 1980s. Human sociobiology was a precursor discipline of evolutionary psychology, and shared with it a commitment to seeking Darwinian explanations of human behaviour. A series of acrimonious exchanges took place between E. O. Wilson, whose 1975 book *Sociobiology* founded the field, and his Harvard colleagues Richard Lewontin and Stephen Jay Gould. The dispute arose from Wilson's claim that many human social behaviours, including aggression, rape, and xenophobia, had a genetic basis, and were adaptations favoured by natural selection because they enhanced the reproductive success of our ancestors.

Sociobiology attracted a variety of criticisms, some of which were strictly scientific. Critics pointed out that sociobiological hypotheses were hard to test so should be regarded as conjectures not established truths, and that cultural influences on human behaviour should not be downplayed. But others objected more fundamentally, claiming that the whole sociobiological enterprise was ideologically suspect. They saw it as an attempt to excuse anti-social behaviour, usually by men, or to argue for the inevitability of certain social arrangements. By arguing that rape, for example, has a genetic component and has arisen by Darwinian selection, sociobiologists seemed to be implying that it was 'natural' and thus that rapists were not responsible for their actions—they were obeying their genetic impulses. In short, critics charged that sociobiology was a value-laden science, and the values it was laden with were very dubious.

In many ways, modern evolutionary psychology represents an improvement over the sociobiology of the 1970s and 1980s. The best work in evolutionary psychology has a strong empirical basis and meets the strictest scientific standards. The naive genetic determinism of the early sociobiologists has given way to a more nuanced picture in which cultural factors, as well as genes, are acknowledged to affect behaviour, and in which cross-cultural diversity is not ignored. However evolutionary psychology continues to attract critics, in part because it shares with its predecessor an emphasis on the 'darker' side of human nature, a focus on matters to do with sex, mating, and marriage, and on the supposed innate psychological differences between men and women. These foci are somewhat surprising, given that human psychology encompasses much more than this. Thus the charge that evolutionary psychology is serving to reinforce existing stereotypes, if only inadvertently, is hard to completely avoid.

One possible response to this charge is to insist on the distinction between facts and values. Consider the suggestion made by some evolutionary psychologists that marital infidelity, or 'extra-pair copulation', is an evolved strategy that human females use to obtain genetic benefits for their offspring when their long-term mate is of low genetic quality. Whether this is true is presumably a question of scientific fact, though not an easy one to answer. But facts are one thing, values another. Even if extra-pair copulation is an evolutionary adaptation, that does not make it morally right. So there is nothing ideologically suspect about evolutionary psychology, despite its rather selective research foci.

Like all sciences, it is simply trying to tell us the facts about the world. Sometimes the facts are disturbing, but we must learn to live with them.

Our second example of possible value-ladenness comes from psychiatry, the branch of medicine that treats mental disorders such as depression, schizophrenia, and anorexia. There is an ongoing debate among psychiatrists and philosophers over how the concept of mental disorder (or mental illness) should be understood. One camp embraces the 'medical model', which says that it is a fully objective matter whether something is a mental disorder or not; no value judgements are involved. Mental and physical disorders are alike in this respect, it is argued. If you suffer from diabetes or emphysema, for example, then your physical body is not working properly; similarly, if you suffer from depression or schizophrenia, then your mind is not working properly. So the boundary between mental health and illness is just as objective as the boundary between physical health and illness, on the medical model.

An alternative view regards mental disorder as an inherently normative category, involving implicit or explicit value judgements. Something gets labelled a mental disorder, on this view, if it involves behaviour patterns that deviate from society's expectations, or that others regard as 'deviant'. For example, homosexuality was regarded as a mental disorder in Western countries until quite recently; it was only in 1973 that the American Psychiatric Association removed homosexuality from the DSM (*Diagnostic and Statistical Manual of Mental Disorders*), and not all of its members agreed. Moreover, medical anthropologists have documented considerable cross-cultural variation in the mental disorders that a society recognizes, something that the DSM has long struggled to handle. So the view that mental disorder is a value-laden or normative concept is certainly plausible. Proponents of this view typically argue that mental disorder is not a genuine medical category at all, but rather an instrument of social control. A radical version of this argument was made by the American psychiatrist Thomas Szasz in a famous 1961 book entitled *The Myth of Mental Illness.*

The debate between the 'medical model' and the view of mental disorder as inherently value-laden is complex. One issue concerns the relation between mind and brain. A point in favour of the medical model is that at least some mental disorders are known to have a neural or neurochemical basis, i.e. they are brain disorders, often arising from faulty brain circuitry. This is increasingly the view of mainstream psychiatry. Since the brain is part of the physical body, this suggests that there is no sharp dichotomy between mental and physical disorders. So if the category of physical disorder is agreed to be objective rather than value-laden, surely the same must be true of mental disorder?

Though powerful, this argument is not conclusive for two reasons. First, for some mental disorders, such as the childhood illnesses autism and ADHD, there are ongoing disagreements over whether they are single, unified disorders at all. These disorders are characterized by clusters of symptoms which often but not always co-occur, with substantial variation from child to child. (This is why autism is called a 'spectrum disorder'.) Moreover, many of these symptoms are found to some degree or other in 'normal' children, who do not meet the diagnostic threshold. This suggests that there is an element of conventionality or arbitrariness in what gets counted as a mental disorder; so even allowing that mental functioning depends on brain wiring and brain chemistry, it does not follow that mental disorder is necessarily as objective a category as that of physical disorder.

Secondly, not all parties agree that physical disorder *is* an objective category. Some philosophers argue that any talk of disorder or illness, whether physical or mental, is inherently normative and value-laden. If someone suffers from a physical disorder this means that their body, or part of it, is malfunctioning—it is not working as it should. But this 'should' indicates a normative dimension, it is argued. Who decides how the physical body 'should' be working? After all human physiology exhibits considerable variation. Some people have 20/20 vision, others slightly less, and others substantially less. Surely any attempt to draw a line and say that this is how human eyes 'should' work involves value judgements? In a society where visual acuity mattered less, for example, the line would probably be drawn somewhere else. So on this view, mental and physical disorder are both value-laden categories.

Against this, other philosophers have tried to bolster the medical model by suggesting that the normativity here is only apparent. Talk of how the body, or the mind, 'should' work can be grounded in a fully objective way, via the concept of biological function, they argue. To understand this suggestion, consider the human heart. The heart pumps blood around the body, and it also makes a regular thumping sound; however only the former is the heart's function—the thumping sound is just a side effect. According to a widely held view, this distinction between function and side effect has an objective basis in facts about evolutionary history. It is because they pump blood, not because they make a thumping sound, that hearts were favoured by natural selection, so exist today. Therefore if a person's heart does not pump blood, then in a fully objective sense it is malfunctioning. When doctors talk about 'heart disease', they are not making any value judgements but simply appealing to what the heart is meant to do, in the sense of its evolved biological function.

A similar story can be told about mental disorders, it is argued. The brain and its sub-components have biological functions; when a person's brain does not perform its function properly, this leads to mental disorder. So in classifying conditions such as schizophrenia and depression as mental disorders, we are not making value judgements but simply appealing to the fact that in patients with these conditions, some part of their brain is not performing its evolved function properly. So the boundary between mental disorder and mental health can in principle be drawn in a fully objective way, via the notion of biological function. In this way proponents of the medical model hope to show that what counts as a mental disorder is not a reflection of prevailing social norms, but rather has an objective biological basis. However this line of argument is controversial, as it rests on assumptions about our evolutionary history that may not be true. For this and other reasons, not all psychiatrists and philosophers accept it.

Finally, note that our two examples of (alleged) value-ladenness in science are of different sorts. In the evolutionary psychology case, the suggestion was that the particular hypotheses researchers choose to investigate, and the answers to them that they propose, serve to reinforce existing stereotypes. If this is true, then in principle it could be remedied by suitably modifying the content of the science, taking care to exclude any possible biases, and applying stricter scientific standards. In the psychiatry case, the suggestion was that the category of mental disorder itself is value-laden, involving implicit value judgements. If this is true, then it is less clear how it could be remedied, if at all, for mental disorder is a fundamental notion in psychiatry. So the value-ladenness in this case is potentially more deep-seated.

To conclude, it is inevitable that the scientific enterprise should find itself subject to criticism from a variety of sources, despite the clear benefits it has brought to humanity. It is also a good thing, for uncritical acceptance of everything that scientists say and do would be unhealthy and dogmatic. Philosophical reflection on the criticisms levelled against science may not produce any final answers, but it can help to isolate the key issues and encourage a rational, balanced discussion of them.

Further reading

Chapter 1: What is science?

A good discussion of the scientific revolution is Steven Shapin, *The Scientific Revolution* (University of Chicago Press, 1998). Detailed treatment of topics in the history of science can be found in J. L. Heilbron (ed.), *The Oxford Companion to the History of Modern Science* (Oxford University Press, 2003). There are many good introductions to philosophy of science, including Alexander Rosenberg, *The Philosophy of Science* (Routledge, 2011) and Peter Godfrey-Smith, *Theory and Reality* (University of Chicago Press, 2003). An excellent collection of papers on general philosophy of science, with extensive commentaries by the editors, is Martin Curd, J. A. Cover, and Christopher Pincock (eds), *Philosophy of Science: The Central Issues* (W. W. Norton, 2012). Popper's attempt to demarcate science from pseudo-science can be found in his *Conjectures and Refutations* (Routledge, 1963). A good discussion of Popper's demarcation criterion is in Donald Gillies, *Philosophy of Science in the 20th Century* (Blackwell, 1993). A good introduction to Popper's philosophy is Stephen Thornton's article 'Karl Popper', in Edward N. Zalta (ed.), *The Stanford Encyclopedia of Philosophy*, <http://plato.stanford.edu/archives/sum2014/entries/popper/>.

Chapter 2: Scientific inference

A clear discussion of induction and scientific inference is Wesley Salmon, *The Foundations of Scientific Inference* (University of Pittsburgh Press, 1967). David Hume's reflections on induction can be found in Book IV, Section 4 of his *Enquiry Concerning Human Understanding*, ed. L. A. Selby-Bigge (Clarendon Press, 1966). A detailed treatment of inference to the best explanation is Peter Lipton, *Inference to the Best Explanation* (Routledge, 2004). The literature on causal inference spans philosophy, statistics, and computer science. An ambitious work on this topic is Peter Spirtes, Clark Glymour, and Richard Scheines, *Causation, Prediction and Search* (MIT Press, 2001). On randomized controlled trials, see John Worrall, 'Why there is no cause to randomize', *British Journal for the Philosophy of Science* 58 (2007), 451–88, and Nancy Cartwright, 'What are randomized controlled trials good for?', *Philosophical Studies* 147 (2010), 59–70. A good treatment of probability and induction is Ian Hacking, *An Introduction to Probability and Inductive Logic* (Cambridge University Press, 2001). The Bayesian approach to scientific inference is expounded by Colin Howson and Peter Urbach, *Scientific Reasoning: The Bayesian Approach* (Open Court, 2006).

Chapter 3: Explanation in science

Hempel's original presentation of the covering law model is in *Aspects of Scientific Explanation* (Free Press, 1965). A useful account of the debate instigated by Hempel's work is Wesley Salmon, *Four Decades of Scientific Explanation* (University of Minnesota Press, 1989). A detailed recent treatment of scientific explanation, with an extensive bibliography, is James Woodward's article 'Scientific explanation', in Edward N. Zalta (ed.), *The Stanford Encyclopedia of Philosophy* (Winter 2014 edition), <http://plato.stanford.edu/archives/win2014/entries/scientific-explanation/>. The suggestion that consciousness can never be explained scientifically is found in Colin McGinn, *Problems of Consciousness* (Blackwell, 1991). The idea that multiple realization accounts for the autonomy of the higher-level sciences is developed by Jerry Fodor, 'Special Sciences', *Synthese* 28 (1974), 97–115. Further discussion of reductionism is found in section 8 of Martin Curd, J. A. Cover, and Christopher Pincock (eds), *Philosophy of Science* (W. W. Norton, 2012).

Chapter 4: Realism and anti-realism

A detailed analysis of scientific realism, with an extensive bibliography, is Anjan Chakravartty's article 'Scientific realism', in Edward N. Zalta (ed.), *The Stanford Encyclopedia of Philosophy* (Spring 2014 edition), <http://plato.stanford.edu/archives/spr2014/entries/scientific-realism/>. Bas van Fraassen's influential defence of anti-realism is in *The Scientific Image* (Oxford University Press, 1980). Critical discussions of van Fraassen's work can be found in Clifford Hooker and Paul Churchland (eds), *Images of Science* (University of Chicago Press, 1985). A book-length defence of scientific realism is Stathis Psillos, *Scientific Realism: How Science Tracks Truth* (Routledge, 1999). The 'no miracles' argument was originally developed by Hilary Putnam; see his *Mathematics, Matter and Method* (Cambridge University Press, 1975), 69ff. A recent analysis is given by Greg Frost-Arnold, 'The no-miracles argument for realism: inference to an unacceptable explanation', *Philosophy of Science* 77 (2010), 35–58. A useful discussion of underdetermination is Kyle Stanford's article 'Underdetermination of scientific theory', in Edward N. Zalta (ed.), *The Stanford Encyclopedia of Philosophy* (Winter 2013 edition), <http://plato.stanford.edu/archives/win2013/entries/scientific-underdetermination/>.

Chapter 5: Scientific change and scientific revolutions

Important papers by the original logical empiricists can be found in Herbert Feigl and May Brodbeck (eds), *Readings in the Philosophy of Science* (Appleton-Century-Crofts, 1953). Critical perspectives on the movement are found in Alan Richardson and Thomas E. Uebel (eds), *The Cambridge Companion to Logical Empiricism* (Cambridge University Press, 2007). Thomas Kuhn's most important work is *The Structure of Scientific Revolutions* (University of Chicago Press, 1963); all post-1970 editions contain Kuhn's Postscript. Kuhn's later thoughts can be found in his books *The Essential Tension* (1977) and *The Road Since Structure* (2000), both published by University of Chicago Press. A good book-length treatment of Kuhn's philosophy is Alexander Bird, *Thomas Kuhn* (Acumen, 2000). Reflections on Kuhn's ideas and legacy are found in Paul Horwich (ed.), *World Changes* (MIT Press, 1993), and Thomas Nickles (ed.), *Thomas Kuhn* (Cambridge University Press, 2002). A useful overview of Kuhn's work, with an extensive bibliography, is Alexander Bird's article 'Thomas Kuhn', in Edward N. Zalta (ed.), *The Stanford Encyclopedia of Philosophy* (Fall 2013 edition), <http://plato.stanford.edu/archives/fall2013/entries/thomas-kuhn/>.

Chapter 6: Philosophical problems in physics, biology, and psychology

The original debate between Leibniz and Newton consists of five papers by Leibniz and five replies by Samuel Clarke, Newton's spokesman. These are reprinted in H. G. Alexander (ed.), *The Leibniz–Clarke Correspondence* (Manchester University Press, 1998). A good discussion of the absolute/relationist controversy is Nick Huggett and Carl Hoefer's 'Absolute and relational theories of space and motion', in Edward N. Zalta (ed.), *The Stanford Encyclopedia of Philosophy* (Spring 2015 edition), <http://plato.stanford.edu/archives/spr2015/entries/spacetime-theories/>. A classic discussion of the species question is by John Maynard Smith, *The Theory of Evolution* (Cambridge University Press, 1993), chapter 13. A useful overview of philosophical work on species is Marc Ereshefsky's article 'Species', in Edward N. Zalta (ed.), *The Stanford Encyclopedia of Philosophy* (Spring 2010 edition), <http://plato.stanford.edu/archives/spr2010/entries/species/>. A historical treatment of the species question is given by John Wilkins, *Species: The History of the Idea* (University of California Press, 2009). Jerry Fodor's original treatment of modularity is in *The Modularity of Mind* (MIT Press, 1983). The extent of mental modularity is debated by Jesse Prinz and Richard Samuels in their contributions to Rob Stainton (ed.), *Contemporary Debates in Cognitive Science* (Blackwell, 2006), 22–56. A useful overview of the modularity issue is Philip Robbins's article 'Modularity of mind', in Edward N. Zalta (ed.), *The Stanford Encyclopedia of Philosophy* (Summer 2015 edition), <http://plato.stanford.edu/archives/sum2015/entries/modularity-mind/>.

Chapter 7: Science and its critics

A book-length study of scientism is by Tom Sorrell, *Scientism* (Routledge, 1991); a useful recent collection is Daniel Robinson and Richard Williams (eds), *Scientism: The New Orthodoxy* (Bloomsbury, 2014). A defence of the view that all genuine questions can be answered by science is given by Alex Rosenberg, *The Atheist's Guide to Reality* (W. W. Norton, 2012). Whether the methods of natural science are applicable to social science is discussed by Martin Hollis, *The Philosophy of Social Science* (Cambridge University Press, 1994). Good treatments of the clash between Darwinism and 'intelligent design' are found in books by Sahotra Sarkar, *Doubting Darwin* (Blackwell,

2007), and Niall Shanks, *God, the Devil and Darwin* (Oxford University Press, 2004). A comprehensive discussion of value-ladenness in science, with an extensive bibliography, is found in Julian Reiss and Jan Sprenger's article 'Scientific objectivity', in Edward N. Zalta (ed.), *The Stanford Encyclopedia of Philosophy* (Fall 2014 edition), <http://plato.stanford.edu/archives/fall2014/entries/scientific-objectivity/>. A good book on this topic is Helen Longino, *Science as Social Knowledge* (Princeton University Press, 1990). The original sociobiology controversy is analysed by Philip Kitcher, *Vaulting Ambition: Sociobiology and the Quest for Human Nature* (MIT Press, 1985). The evolutionary psychology programme is set out in Jerome Barkow, Leda Cosmides, and John Tooby (eds), *The Adapted Mind* (Oxford University Press, 1995). A detailed critique is offered by David Buller, *Adapting Minds* (MIT Press, 2005). A useful overview is Stephen Downes's article 'Evolutionary psychology', in Edward N. Zalta (ed.), *The Stanford Encyclopedia of Philosophy* (Summer 2014 edition), <http://plato.stanford.edu/archives/sum2014/entries/evolutionary-psychology/>. Good discussions of the concept of mental disorder are found in Rachel Cooper, *Psychiatry and Philosophy of Science* (Routledge, 2007); and in Christian Perring's article 'Mental illness', in Edward N. Zalta (ed.), *The Stanford Encyclopedia of Philosophy* (Spring 2010 edition), <http://plato.stanford.edu/archives/spr2010/entries/mental-illness/>.

Index

J

K

L

M

N

O

P

Q

SOCIAL MEDIA
Very Short Introduction

Join our community

www.oup.com/vsi

- Join us online at the official Very Short Introductions **Facebook** page.
- Access the thoughts and musings of our authors with our online **blog**.
- Sign up for our monthly **e-newsletter** to receive information on all new titles publishing that month.
- Browse the full range of Very Short Introductions online.
- Read **extracts** from the Introductions for free.
- If you are a teacher or lecturer you can order inspection copies quickly and simply via our website.

Online Catalogue

A Very Short Introduction

Our online catalogue is designed to make it easy to find your ideal Very Short Introduction. View the entire collection by subject area, watch author videos, read sample chapters, and download reading guides.

http://global.oup.com/uk/academic/general/vsi_list/